Good Words

The content of *Children of the Stars* is excellent. The examples of culture and education woven into the story are both insightful and relevant. As an Anishinaabe and Indigenous person I see many connections with my own culture, and appreciate the form and format. We know that to truly connect with Indigenous people you must reach their hearts before their heads, and I believe Ed Galindo has captured that spirit in this book. WELL DONE! As our mutual friend Mr. GooGoo says, "Let's have tea before we talk." This book is an invitation to have tea, to have a peek at the important world of Indigenous relationships and the role they play in learning. Kinanaskomitin—with great respect.

—Darren McKee
Saskatoon School Board, Saskatchewan, Canada

I felt like Ed Galindo was talking to me.

—Dr. John Herrington
first Native American NASA astronaut

Storytelling is at the heart of how we visit, and revisit again, year after year. The memories in these stories are deeply personal. I've been able to travel back in time, to a time of friends, family, and relatives. Some have since passed and some are still here. Stories are how we live on. These stories are great recollections of honorable people. Aho, All my relations.

Navashay umbunihi.

—Dr. Sammy Matsaw
former NASA student at Sho-Ban High School

CHILDREN OF THE STARS

Children of the Stars

Indigenous Science Education in a Reservation Classroom

ED GALINDO
with LORI LAMBERT

Oregon State University Press   Corvallis

Cataloging-in-publication data is available from the Library of Congress.

ISBN 978-0-87071-201-2 (paperback); ISBN 978-0-87071-202-9 (ebook)

♾This paper meets the requirements of ANSI/NISO Z39.48-1992
(Permanence of Paper).

First published in 2022 by Oregon State University Press
Printed in the United States of America

Oregon State University Press
121 The Valley Library
Corvallis OR 97331-4501
541-737-3166 • fax 541-737-3170
www.osupress.oregonstate.edu

Dedication

SUMMER OF '97
by Autumn Pratt

I dedicate this book to all of the Sho-Ban students
To this day I can recall,
The best summer I've had of all,
The winding rugged roads,
And all the memories that it holds,
Is up to me to behold,
Blue mountains and green trees,
All around as far as the eye can see,
And then the hidden dusk comes crawling in,
The day is almost at the day's end,
Hamburgers and hotdogs on an open grill,
Soon our hunger is fulfilled,
Around the fire everyone gathers,
Laughing and talking seems like for hours,
Rain, lightning, and thunder says it's time to rest,
Away in our tents we are nodding off,
The rain and the thunder's story is the last we hear,
We fall asleep but not for long,
Soon the birds awake and sing their morning song,
Then in the distance I hear "first wakeup call,"
And just then a new day filled with freedom starts for all.
Thank You.

Contents

Thank You

The Shoshone-Bannock tribe: You gave me the freedom to work with your most important and valuable resources in the tribe, your Sho-Ban students.

The "three wise men" of the Shoshone-Bannock Reservation: Mr. Ernest Wahtomy, Mr. Lavern Broncho Sr. (walked on), and Mr. Kevin Callahan. All three care deeply about the Shoshone-Bannock students.

Linda Jay, my sister! Linda Jay was not only our bus driver, but our student chaperone. She is quite awesome and makes the best Indian tacos around.

The mentors, and there were many. If I left out your name, I apologize.

SFC John Moller (retired), US army reserves, 1016th Quartermaster Company, Petroleum Pipeline, Idaho National Guard, Gowen Field. John was our Shoshone-Bannock High School JROTC instructor. John provided the students the resources, enthusiasm, and care that all young people need and deserve. He is a good friend.

Mr. Tony M Messina. Retired broadcasting engineer. He cannot be thanked enough for his help on many projects. His banjo music is the best!

J. R. Simplot analyzed our samples and provided internships for my students.

Families of my students: I appreciate the trust you gave me to take your student with me across the nation to the many places we visited. Houston, Texas, and Cape Canaveral, Florida, just to name two. Many times, quite far and away from the reservation and family. It was an honor.

I want to say a special thank you to two people who helped to get my story out for others. We, as humans, really do not do anything totally by ourselves on this planet. I owe a lot to these two good humans!

Kim Hogeland, my editor. Whom I met a long time ago in the rolling hills of Kansas. It was my good fortune to get to know you.

I thank you again for your help and patience. Thank you for being a part of my circle.

Dr. Lori Lambert, my coauthor. I cannot say enough words to express my heartfelt thank you for your advice, examples, and understanding of not just words, but the feeling and meaning behind the words and the journey they take us on. I thank you for your *good dreams* and acting on those dreams. I thank you for your never-ending persistence. You became a family member of mine a long time ago, and I thank you again for the journey as it continues . . .

Finally, I want to say a personal thank you to my own family. Without them, all that I am or want to be would not be possible. They have given so much to my life. I view them as a sacred blessing. They know who they are! Thank you all.

In this story, we will use the following terms: American Indian, Native American, Indian, Natives, and the People. All are meant with respect to describe an honorable nation of people on this Mother Earth.

Introducing Ed

Storytelling is a very important aspect of Native America. It is not just the words, and the listening, but the actual living of the story.

—Leroy Little Bear (foreword to Cajete 2000, xii)

Depending on my old memory, I am going to share my true story with you. My story is about my experience as an American Indian science teacher working at an American Indian high school on the Shoshone-Bannock Indian Reservation at Fort Hall, Idaho. This story is also about my personal journey in Indian country. The insights I offer are mine alone. I do not generalize for all tribes, nor do I generalize for all teachers. I will relate stories as they were told and shared with me, and how I remember them. These stories will be told in the American Indian style of sharing, in a circular fashion, unlike the Western style of storytelling, which is linear. As I tell the story, stick with me and the point will come in time. I will not talk about sacred stories that are not meant to be shared, or sacred ceremonies that are done on the reservation. I apologize if the stories are shared out of season. I will protect the identity of the students by not using their real names. I will mention the tribe, as it is common knowledge that the Shoshone-Bannock tribe and I are forever linked together in history and time. I am proud of this fact.

MY FAMILY

Although the focus of this story is on the lives of the students who attended Sho-Ban High School in Idaho, in Indian cultures it is a sign of respect to introduce one's family; doing so gives Indian people a frame of reference of who family members are and where you come from.

I am a Yaqui Indian and was raised with my grandfather and grandmother, who lived with us in the same little houses in Texas

My grandfather, Felipe Galindo, with my father, Mike Galindo, and my Aunt Virginia, circa 1946.

and Idaho. Carl Waldman in *Encyclopedia of North American Tribes* describes the history of my people:

> *The Yaqui or Hiaki or Yoeme are an Uto-Aztecan speaking Indian people. They inhabit the valley of the Rio Yaqui in the Mexican state of Sonora and Southwestern States. They also have communities in Chihuahua and Durango. The Pascua Yaqui are based in Tucson, Arizona. Throughout their history, the Yaqui remained separate from the Aztec and Toltec empires. They were similarly never conquered by the Spanish, defeating successive expeditions of conquistadores in battle. However, they were successfully converted to Christianity by the Jesuits.* (Waldman, 2006, 323–324)

My family's roots extend into the southwestern part of rural Texas, a glorious section of the state, adjacent to the border of the Rio Grande River and the mountainous topography of the Chihuahuan Desert. The area is now called Big Bend National Park. It is the largest protected area of the Chihuahuan Desert ecological system in the United States.

My great-grandfather was the leader of our family. He was a Yaqui Texan rancher who raised sheep, goats, horses, and mules. Even today, it is a challenge to grow crops and raise cattle on the dry desert lands, but my family knew how to live, and they thrived. When America entered World War II, after the Japanese attack on Pearl Harbor, he sold all of his animals and supplies to the United States government for the war effort. As many Indian men and women on reservations and in cities all over America enlisted in the military, my great-grandfather did his part to protect America because it was his homeland. When there was no one to help or care for the people, it was important for him to support the people of his area. Many people reading this story do not realize that, per capita, Indian men and women have the highest numbers of enlistments in the military forces of the United States (National Indian Council on Aging 2019).

Chihuahuan Desert Nature Park. Photo: Patrick Alexander/Flickr.

As my grandmother's first grandson, I was showered with her love early on in my life. She refused to drive, wear pants or long dresses, or speak English. She spoke Spanish and the Yaqui language. She always spoke to me in her Indian language. In the spring of each year, my grandmother gathered her digging sticks and took me to collect medicine plants in the wild meadows and mountains. In her language, she named each plant. She explained what each part of the plant was used for, whether it be the roots, stems, petals, or leaves, and she described how it was to be prepared for human health. For some ailments, the stems were used, for others she prepared the roots, and for other health issues, the leaves or blossoms were used. Some plants were steeped as tea, others as poultices. Sage was dried to make the smudge, and some plants were used as food. Plant knowledge was well-known among the Yaqui tribe, and she was educating me so her knowledge would not be lost. All instruction was in her Indian tongue. She quizzed me on what she said. If I did not pass, she would start again with her instruction. My grandmother was very patient with me.

My father was a Yaqui Indian and my mother was non-Indian. I had a rich upbringing as we all lived together in a multigenerational, multilingual home. From my grandparents I learned respect and how to be patient. Before I was born, my family left Texas and traveled to Idaho in search of work. This was a common practice among the people of that era, when many natural disasters, as well as man-made ones, forced people off the land. For most of the 1930s, people in the plains states suffered through the worst drought in American history, as well as hundreds of severe dust storms. By 1940, 2.5 million people had abandoned their farms in the Dust Bowl and headed West (History.com).

We lived and worked on a cattle ranch in Idaho where my grandfather taught me how to ride and break wild horses and work with the cattle. He also taught me valuable lessons about caring for horses, but more importantly how to respect this animal. Later in life, the lessons of compassion and respect my grandfather taught me would help me with my teaching. I learned how to treat a sick animal, first by observing the animal's normal behavior, and then by observing when there are changes in that behavior. From observing behaviors, I recognized when a horse, cow, or working cow-dog was sick. I was learning veterinary medicine and how to formulate a health plan for sick animals. Learning how to treat sick animals was, and is, common knowledge in all ranching communities. Not only was I taught about the animal world, I also learned about the plant world, medicine plants, and caring for Mother Earth. I learned why a crop failed, or why certain species of wheat and grain grew better in some types of soil than in others. Lastly, I was taught about the human world, and why some people hate and others do not.

My parents taught me how to work for everything I wanted. My childhood was one of joy, as the family in our little ranch house in Idaho included my sister, my grandfather and grandmother, mother and father, and at times my auntie and uncle. I enjoyed the crowd.

MY CAREER PATH

After graduating from high school, I attended the College of Southern Idaho (CSI), a community college, and transferred to the University of Idaho (UI) to complete my bachelor of science degree in animal science and chemistry. Following UI, I earned a master's in health education at Idaho State University (ISU) with an eye to enrolling in medical school.

I never imaged that I would become a science teacher. My plan was to become a medical doctor, and I was well on my way to making this happen. I was accepted into an osteopathic medical program. Part of the medical program involved taking summer medical school prep programs at Ohio University in Athens, Ohio. I was an excellent student and doing well with the medical school classes. I passed all the exams and classes the summer before I was to enter medical school. But I had a nagging question about whether I would really fit into the culture of medical school, as I was just not happy being at Ohio University.

Before being accepted into the medical school program, I worked at a local hospital in Pocatello, Idaho, eight miles away from the Sho-Ban Reservation. I enjoyed my time at the hospital. At the same time, I made several friends from the reservation who invited me to their homes for dinner. The reservation community felt more like home than the medical school community.

My job at the hospital was to fill medical carts with supplies and provide a general helping hand to the hospital staff. My title was "runner," because that is what I did, run around helping where needed. One night, I was helping in the emergency room when a horrific trauma case arrived by ambulance that involved a baby and mom. The mom survived the event, but the baby did not. I tried and tried but could not get that event out of my head. I thought about it a lot. A trauma nurse, who was watching me struggle, took me aside and told me that, if I was going to be a physician, I would have to get used to situations like that as I would be facing similar situations for the rest of my life. I decided then that being a doctor was not for me. I was twenty-seven years old. I reexamined my life's goals and decided to be a science teacher instead of attending medical school. I thought more about my dream of being a science teacher on the reservation. I was told by an Idaho State education professional that being a science teacher was good career, but he advised me not to teach on the reservation, since the only school on the rez at that time was an all-Indian alternative school and teaching there would ruin my career. Instead of discouraging me, all this talk focused me on the need to do more to become a teacher on the reservation and work hard with Indian students!

As I look back on my life, I am not unhappy with the path I decided to travel, teaching versus medicine. Most of my teaching life has been with Indian students and communities.

NASA, MY PhD, AND THE ROLE OF THE STUDENTS

At a critical point in my career as a science teacher at Sho-Ban High School, I was offered a chance to travel to Utah State University (USU) in Logan, Utah, and earn a PhD as a NASA Fellow through the Utah Space Grant. Being a NASA Fellow meant that NASA dollars paid for the education, but NASA wanted something in return. My first responsibility was to be a top student and meet the extraordinarily high GPA requirement. While I was a graduate fellow with NASA, I was expected to concentrate on my PhD and not work. I focused my PhD on education and engineering with a physics emphasis. It was one of the first interdisciplinary programs at USU. All of my time was spent learning, researching, and passing exams with the highest marks. It was fun but exhausting. I also agreed, as a NASA Fellow, to build and fly a scientific instrument in space, aboard a space shuttle.

Both undertakings were arduous and challenging projects. I elected to include my high school science class in the NASA research component. I informed NASA that the education of the youth was everyone's job, not just the Department of Education. Because of my diligent academic work and my idea for my experiment on the space shuttle, I was granted my NASA Fellowship with USU, and was on my journey to earn a PhD.

My dissertation at USU reflected research in two fields—research and education—and encompassed two sections. The first section of research involved the required NASA space experiment. The second portion was the required research in education. All research begins with a hunch or a hypothesis. My hypothesis for the education research was that students who stayed in school or left school made those decisions with encouragement or lack of encouragement from their families. The research question for the education portion was "Why do Indian students have a high dropout rate or low completion rate compared to a national average?" The hypothesis for NASA research was that I could build experiments that would fly in space with the help of my science students on the reservation. It would fly; the students would learn a lot and be excited about learning science. My research question with NASA was "How does water react in space, and would it mix with solids?"

Since I wanted my students on the reservation to help with my experiment, we formed a science club. We had to obtain permission from NASA to use their initials. I received permission from the director himself. NASA: the Native American Science Association—we were on our way. After intense discussions, my high school science students and I decided we wanted to grow food and "paint" in space. Both are complex issues, and many years later, both continue to interest me. The students worked on sections of the experiment where they had the skills. Although lacking advanced electronic skills, they applied their excellent critical thinking skills, and could ask pertinent questions and work with mentors to complete the technical aspects. We designed and flew the first Indian experiment in space, which orbited the Earth aboard a space shuttle. It was *damn hard* to do but a great learning experiment for me and the students. As part of my NASA experiment, we led the first Indian *Vomit Comet* team in microgravity aboard an aircraft (KC-135) that flew out of Houston, Texas.

THE SHO-BAN STUDENTS

My Shoshone-Bannock students taught me many, many things. For example, where to find my old pickup after they secretly and jokingly hid it during school hours and basketball practice. They also taught me how to teach effectively, so they could learn. The techniques I learned in the classes at the college of education were not effective in my reservation classes. My students demanded to know why they needed to learn things like photosynthesis. Not just how photosynthesis worked in green plants, but a higher-order *why*. This made sense to me. I learned to "sell" my education idea in about thirty to sixty seconds, or it did not make valid sense to them. This is a valuable lesson. I owe everything to my Shoshone-Bannock students, as they taught me to be a teacher of what they needed, not what someone off the reservation thought they needed to know. They knew what they needed to know. I learned from them that teaching is not for the faint of heart. It has to come from the heart. The students will know when it does!

As a science teacher on the Shoshone-Bannock Indian Reservation and a Yaqui Indian, I was an outsider. It was important that I earn trust, respect, and admiration from the students and the reservation community. Earning trust is *not* a given. Students needed time to know me and test my heart about the subject matter and themselves. This takes time and tests patience. I believed that if I was strong and true of heart, then I could help Indian students, whose families understand traditional ecological knowledge (TEK) and appreciate its relationship with western science. For example, climate scientists today are partnering with Indian Elders, ecologists, and other Indian scientists to understand traditional ecological knowledges in the fight against a warming planet.

There are many heroes in this story and many, many problems that had to be solved before the students could fly a science experiment in space with NASA. This is a story of how our team learned from and shared with each other. This is a story about some of the brightest stars I know in our solar system, the young people I learned from and taught on the Sho-Ban Reservation. This story is shared with honor, respect, and compassion and is dedicated to these students, some of whom are now with the stars. May their spirits forever fly high and free. Gather 'round. The fire is ready. Come and hear our story!

Introducing Sho-Ban

This is a story about the beauty of a reservation. This is a story about the beauty of the landscape, the beauty of reservation people, and the beauty of Indian students. This is a story about determination, bravery, humor, science, compassion, sorrow, hope, space, and stars. Part of this story is about working, learning, and teaching the most important natural resource of any tribe, their young people. I am going to share my story of what I learned about the Shoshone-Bannock tribes from the view of an outsider. Being an outsider is comparable to living in two worlds. As a Yaqui Indian, from a tribe located in the Southwest, I have been honored to be a guest of the Shoshone-Bannock people for over thirty-five years. The stories I will share about the people were the ones told to me by the Shoshone-Bannock members themselves. With respect to the tribe, and compassion for their feelings, I will share what I know about the Shoshone-Bannock people.

Originally, the Shoshone and Bannock were two different tribes with two different languages. The Shoshone are direct descendants of an ancient people who called themselves Newe (nu-wee), which means "The People." Possibly because of food shortages or other environmental challenges, the Shoshone separated into three main groups—Northern, Western, and Eastern—and settled in areas of California, Utah, Nevada, Wyoming, and Idaho. The largest group settled in the Snake River Valley, Idaho, and are sometimes referred to as "Snake Indians." Shoshone Falls, the Portneuf (Port-noof) River, and the city of Pocatello are named after Shoshone chiefs (Arnold et al. 2010).

The Bannock Indians are a Shoshonean tribe who long lived in the Great Basin in what is now southeastern Oregon and southern Idaho. Calling themselves the Panati, they speak the Northern Paiute language and are closely related to the Northern Paiute people, so much so that some anthropologists consider the Bannock to be simply one of the northernmost bands of the Northern Paiute. Early on, and even today, the Bannock subsisted primarily on fish and small game. They fished with harpoons, hand nets, and weirs built from woven willows (*History of the Eastern Shoshone* 2015).

I was told that back in the day, the tribes traveled in small groups and mixed with one another on hunting trips. Over the years, the people from each tribe eventually intermarried and became known as the Shoshone-Bannocks. I learned, growing up in Indian communities, that before contact with Europeans, the Indian concept of life was simple and at the same time very complex. For example, Indian people sought a balanced life, taking only what they needed for that day to feed the family, or what they needed to preserve for winter. The traditional Shoshone people gathered wild rice along the riverbanks, and in the forest they foraged for pine nuts, seeds, berries, nuts, and roots. Families gathered eggs from nests, if they could find them. Meat was also a very important item in their diet. They hunted big game animals such as deer, buffalo, elk, moose, and pronghorn (*History of the Eastern Shoshone* 2015).

Before the coming of the white man, food, shelter, and clothing were the only requirements that Indian people needed. They camped in different areas throughout the season to gather, prepare, and dry their food for winter use. Men hunted or fished to provide food for the tribe. They hunted wild game including deer, elk, moose, bighorn sheep, and buffalo. The animals were important because their meat, bones, and hides provided tools, clothing, shelter, and nutrition.

Food preservation was important as the winters were long and the summers short. Then, as now, meat was dried or smoked. Vegetables and berries were dried, smoked, or pickled. Men taught young boys to hunt, and how to create and care for their hunting weapons. Many women were Indian healers, and they often spent their days away from encampments teaching young girls to gather medicine plants or to weave their famous baskets. Salmon was an important food to the Shoshone-Bannocks, as it is today, and they cherished their close relationships with the earth and the animals:

> *These animal relationships were expressed through ceremonial relationships that focused on their connection and their ability to connect humans to the universal order. The world and animal renewal ceremonies, practiced by all tribes, expressed the human*

> *responsibility to preserve, protect, and perpetuate all life. . . . The Salmon ceremony received the most elaborate rites, though it varied from place to place.* (Cajete 1991, 104)

Before the arrival of strangers from Europe, Indian people believed that the Creator put everything on this earth for a purpose. Nothing of Mother Nature was for sale. Their belief was that land was there for everyone to use, not abuse. They believed that everyone was equal, with each person having a place within the tribe. Indians could not understand the mind-set of non-Indians who placed a monetary value on things produced by Mother Nature (Idaho Centennial Commission 1990).

There are three main traditions of the Shoshone Indians: the Vision Quest, the Power of the Shaman, and the Sun Dance. There is a great deal of focus put into the supernatural world. The Shoshone Indians believe that supernatural powers are acquired through vision quests and dreams (Arnold et al. 2010). Over time, these ceremonies may have been adopted by the Bannock tribe as well.

As with many reservations, the Fort Hall Reservation, also called the Shoshone-Bannock Reservation (the name used throughout this book), was the result of the 1868 Fort Bridger Treaty between the American government and the tribe, in which the Sho-Ban tribe was forced to relinquish large areas of their aboriginal homelands. The Shoshone and Bannock Indians once roamed over vast stretches of mountains and sagebrush-covered plains of what now are Idaho, Utah, Nevada, western Montana, and Wyoming. Theirs was a land of creeks, rivers, hot and cold springs, rugged mountains, valleys, and meadows (Heady 1973; Broncho 1995).

The Shoshone-Bannock Reservation included some 1,800,000 acres and approximately 4,500 residents. In exchange for the land given up by the tribe, the government promised to provide protection from settlers, who did not respect the Indians. The government also promised to supply the tribe with education, health care, food, and white man's clothing, such as wool blankets. Within a year of the treaty signing, the government broke those promises. Settlers and their livestock began to infringe on the tribe's natural resources, which were meant for food use by the tribe. The thousands of settler people traveling by wagon trains to Oregon and California also impacted the tribes' ability to hunt and gather. The government's broken promises led to a war for Indian survival. Ultimately, Fort Hall was founded in 1868 as a trading post for trappers (*True West* 2021) and later on for travelers heading to Oregon and California (Shoshone-Bannock Tribes, www.sbtribes.com).

Two decades after the borders of the Sho-Ban Reservation were established, Grover Cleveland, then president of the United States, determined that Indian people would benefit from white American settlers living adjacent to them on reservations. Cleveland believed that Indian people would learn to be "white" farmers. In 1887 he signed the Dawes Act, also known as the General Allotment Act, into law. Traditionally, the tribes held their land communally, but the law allowed the president to break up reservation land into "lots" of small allotments to be parceled out to individual Indian members. Generally each adult Indian member was allotted 80 acres. Couples were "given" 160 acres. Children under eighteen years of age were "granted" 40 acres. The law was designed to bring non-Indian farmers and ranchers into the reservation by giving away "free land" that was not assigned to Indian members. However, the hidden agenda was to assimilate Indians into mainstream US society by annihilating their cultural and social traditions. The Dawes Act "reduced the overall land base by half and furthered Indian impoverishment and United States [government] control" (Dunbar-Ortiz 2014, 158).

Eventually, the Dawes Act allowed the government to strip away over 90 million acres of Indian land from Indians, then sell that land to non-Indian US citizens. As a result, the Sho-Ban Reservation was divided up into allotments, resulting in loss of land for the tribe. Between 1888 and 1902, the Shoshone-Bannock were pressured to cede away nearly a third of the reservation to the growing town of Pocatello. Between 1911 and 1913, half a million acres were allotted to Shoshone-Bannock tribe members. The remaining six hundred thousand acres were subsequently sold to non-Indians for a fraction of their worth (Larsen 2019, 1). Eventually, "the tribes were forced to cede thousands of Indian acres for all kinds of things that non-Indians wanted, including a railroad, a town and, later on, an airport, built with land given by the tribe in good faith during World War II, with the understanding the land would be given back to the tribe when the war was over. Another broken promise" (Slapin and Seale 1998, 349–350).

Even then, under the guise of the federal Bureau of Indian Affairs (BIA), the government was in control of the reservation and the People. The government had no understanding of the Peoples' culture or their needs for traditional foods. Their agents on the reservation made harmful decisions. For example,

> *Horses symbolized power, mobility, and strength to the Shoshone-Bannocks, who required at least ten steers as payment for just one horse. To the government, the horses were just obstacles in the way of progress. The final straw was in September 1920, when government agents quietly rounded up 1,200 horses and sold them for less than $6 apiece—to be turned into chicken feed.* (Larsen 2020, 1)

US citizenship was granted to Indians in 1924. However, only Indians who accepted the division of Indian lands were allowed to become US citizens.

Today, the reservation encompasses 540,764 acres of land. It is a land laced with clear running creeks, mountains, meadows, and unbelievably clear springs. It is a true paradise. Reservations in the United States are located in some of the most beautiful locations I have ever seen. Most reservation lands are remote and the landscapes are undisturbed—meaning that they are not to be disturbed because they are sacred to the people. In the modern day, we call this "sacred geology."

Reservation lands are not all the same, just as not all people are the same. Some reservations are large, including lands extending to several states. Others are small, consisting of a few acres of land. The land is special for Indians for many different reasons. The People believe that the Creator placed them on the earth to care for the land. Reservations are home, and exhibit a sense of place like anyone's home. Countless reservations have consecrated sites where ceremonies are still performed today as they were for thousands of years. One can visit some reservations and still view evidence of early arbors from the sacred Sun Dance, or ancient rocks that held down the tipis, even evidence of buffalo jumps. All reservation lands hold the bones and memories of loved ones, ancestors walking in the spirit world.

Reservations, like any communities, have their fair share of problems. Many Indian members suffer from health disparities resulting from extreme poverty, high unemployment rates, alcoholism, suicide, and drug abuse. Nevertheless, reservations also have positives, such as a sense of family, sharing, humor, and love not found anywhere else. As in any community, not all people living and working on reservations have problems. I know many Indian people on many reservations who work hard, take care of their families, and help others the best they can. Their lives are similar to thousands of families around the world. They go about the business of making a living with a quiet determination and dignity, caring deeply for their tribe, family, and planet.

The Shoshone and Bannock tribes learned from nature how to survive in, at times, a very hostile environment. The learning curve was quite high. If one was unable to learn from and respect nature, survival might be in question. The people became skilled in the art of medicine-plant healing, navigation by land and stars, and science and mathematics. Mathematics and science were the keys to building dwellings and crafting clothing from the resources found in the local environment. The people understood the chemical processes for combining dried meat and berries to make pemmican. Both men and women became skilled at drying and smoking meat for the long winter months when hunters could be stranded at home because of deep snow and subzero temperatures. The people were taught by the elders and other teachers, who did not give "F" grades. Young people either passed with the knowledge they needed, or they tried again and again with their mentor and learned what skills they needed. The final test was to demonstrate to the mentor that they could do the task and live. For example, making fire. If they were unable to start a fire, they tried until they could.

The tribe depended on everyone accomplishing their work and doing it well. The people learned much more than survival; they learned how to live in harmony with Mother Earth. They learned how to take care of one another. This is now called Traditional Knowledge, although I like to think of it as Timeless Knowledge. "Traditional" implies a specific tradition and asks the question which/whose tradition? "Timeless" is the knowledge that was and is known for all time. For example, the knowledge of water as a medicine. All people and plants need this medicine called water. All life on Mother Earth needs clean water, not just those who can afford to buy it. Water is a critical gift for everyone. More importantly, the tribes learned to appreciate the great mystery of life. They showed appreciation and respect for water and the land, with ceremonies of prayer, dance, song. For great festivals of singing, dancing, learning, family time, and storytelling, the two tribes often camped and hunted together, or gathered in an area that was good for camping, like the Snake River bottoms of Idaho.

Unlike eastern tribes who had been removed to the West by President Jackson's Removal Policy of 1830 (like the Eastern Band of Cherokee and the East Coast tribes impacted five hundred years ago by the Jesuits), most Indian tribes in the western part of North America had no formal written language. However, they recorded visual stories on wampum belts, buckskin, "newspaper rocks," and other surfaces. Storytelling in Indian languages was and is a form of entertainment, learning, enlightenment, and teaching.

They knew the importance of passing on knowledge that was key to the survival of the next seven generations. When stories are told in Indian languages, the words are powerful and descriptive. However, when translated into English the deep meaning of the Indian language is lost, and the richness of the story is not honored. Storytellers were men and women highly trained to teach the knowledge in a very specific storytelling way. It was and still is an important job within tribes. Storytellers are trained to weave history, science, geography, mathematics, medicine, and the entire Indian curriculum into stories—at times amusing—with some hidden lessons to think about. Even today, storytellers relate many kinds of stories. They tell sacred and spiritual stories of creation. Historical stories tell of Indian people and describe where they came from, or relate the exploits of a particular hero. There are first-person stories and stories of current events. When an elder storyteller dies, the elder's library of stories may die with her. Some people today call this holistic learning. I like to think of the storytellers as the walking librarians and historians of the tribe.

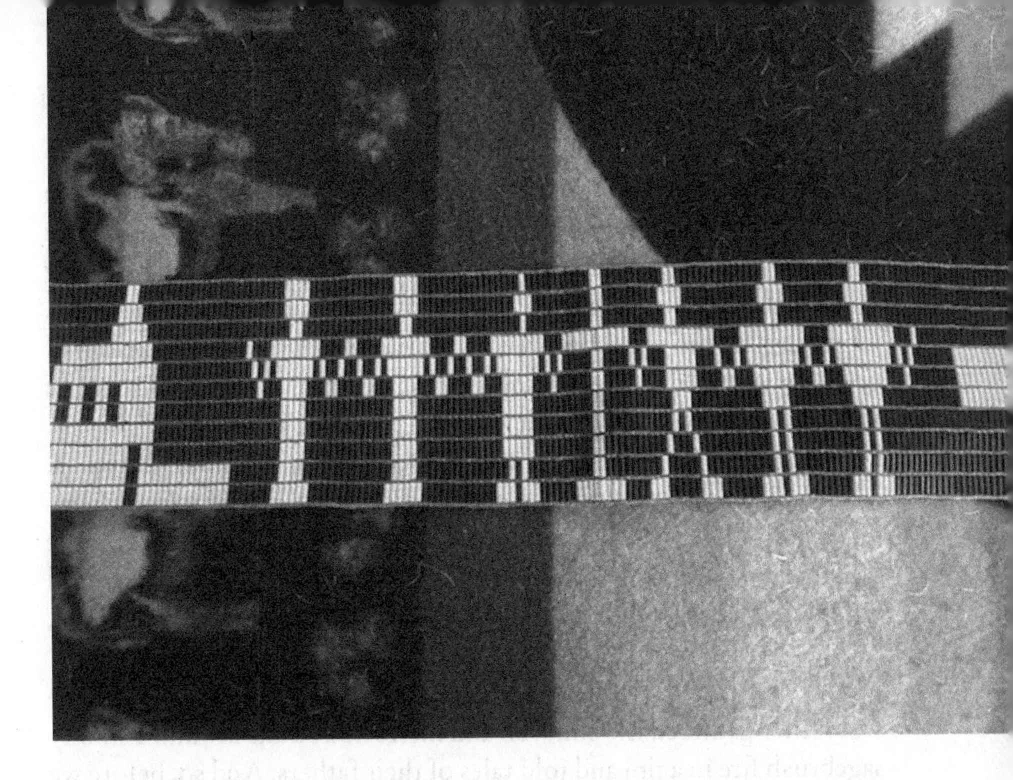

(*Above*) Story told on wampum belt; (*below*) story told on buckskin. Photos: Lori Lambert.

As with many things to be learned or talked about with Indian people, I will continue with a story. Stories are the origin of Indian oral tradition and are the means for sharing knowledge and passing it onto the next generations. Indian stories are older than we are. Countless stories are thousands of years old and carry the wisdom of all those years. It is through the culture, history, and the landscape of the tribe that the story is understood.

The following story is told with respect for the Shoshone-Bannock. There are special rules regarding the telling of stories. Stories are generally told in winter, and there are times when stories should not be told, usually in summer. So, as you read this story, know that it is told with deep respect and if you read this and it is not in the right season, prayers have been said with respect for this correction.

As the stories were told long, long ago, on long winter days, the storyteller gathered the children and interested adults around a smoky sagebrush fire in a tipi and told tales of their fathers. And so, before we tell you about "children of the stars," let's read a story about how the Earth was created. Let's see what we can learn. Our fire is ready, come and share! Have patience with this important story and time as told to me by a Sho-Ban Elder. The lessons he taught me are still important.

HOW THE WORLD WAS MADE (AS TOLD TO ME)

I was told this story around a campfire. It was late fall, and the first snow of the season had fallen. The air was crisp and clean. The storyteller, my teacher, was an elder of the tribe and had one of the most delightful smiles and the most infectious laughter that I ever heard. He had a gentle spirit, and I called him my friend. This evening, he wore a single grey and silver braid down his back and faded Levi's with a cowboy shirt, boots, and an old army jacket. This story has multiple meanings at many different levels. Your job is to pay attention and wonder about the story. Do not overthink it.

Long, long ago before most things were made, there was water everywhere. Not one bit of dry land showed above the huge sea that covered everything. Now Apa, the Great Father of all, lived way up in the clouds and stars, away from the water. He had three sons, water children, who lived in the water, floating upon its surface. One of the sons was Hanee, the beaver, another was Tindui, the otter, and the third was Bammboka, the muskrat.

One day Apa began to wonder what could be under all that water. If we could just have something besides water, he thought, then perhaps many other kinds of creatures could live there. We could have animals that would not have to live in the water.

So Apa left his home in the stars and clouds and went down to the water to visit his sons. "Children," he said, "I want you to help me find out what is under all this water."

"How can we do that?" asked Hanee. "I don't think there's a bottom to the water." And he slapped his tail on the surface, making an echoing noise.

"My brother Hanee is right," agreed the otter. "There is so much water here that no one could find the bottom." And Tindui swam around in a huge circle to show how big the water seemed to him.

"Maybe my brothers are right," agreed Bambooka. Then he, too, swam around in a circle. But when the muskrat returned to the others, he looked very wise. "Maybe," he said, "just maybe, we could dive down far enough to find what is at the bottom of the water."

"What if there isn't any bottom?" asked Hanee.

"We could try," said Tindui.

Apa listened as his three sons argued. Then he said, "Banbooka has a good idea. I'll send you down one at a time to find what is under the water. Who would like to try first?"

"Oh, let me go," shouted Hanee. "I am a very good swimmer, if it is possible to dive to the underside of the water, I'm the one who can do it."

"Very well. You may go, but be sure you don't stay down too long," warned Apa.

So, the beaver dived, while up on the surface his father and brothers waited. Hours passed. Finally, after a very long time, Hanee floated up, gasping for breath.

"What happened?" asked Apa. "Did you find the bottom?"

"I met the fish and other creatures that live in the water. I spent so much time talking with them that I ran out of air," gasped Hanee.

"But did you find the bottom?" asked Bambooka.

"No. I had to come up before I got down far enough," admitted the beaver.

"Let me go Father," urged Tindui. "I'll dive down and find the bottom if anyone can."

So, the otter started down into the water. Around and around he swam, meeting the fish and having a delightful time.

Upon the surface, Apa, Bambooka, and Hanee waited. Finally, after a very long time, the exhausted otter came to the surface, gasping for breath.

"I didn't find anything," he said.

"Please, Apa, let me try," pleaded the muskrat.

Hanee and Tindui laughed. "Little Bambooka cannot make it," they said.

"He is small," said Hanee the beaver.

"Yes," said Tindui, the otter. "If my brother, Hanee, and I could not find the bottom, how can little Bambooka?" said Tindui.

"It cannot be done," said Hanee, the beaver.

"Please, Apa. Let me try," said Bambooka. "I know I can do it!" said the muskrat. The two brothers, Tindui and Hanee. started to object.

"Very well," agreed Apa. "But if you do not find what is underneath, we must be content with a world of water."

Bambooka was excited. He knew he could do it. He was small, but his heart was large.

So Bambooka dived. Straight as a bowshot arrow he went. He paid no attention to the fancy fish who called out to him along the way.

Down, down, down he dived. He was about out of air, but still he dived. After a long time Bambooka surfaced! On his nose he held a ball of mud!

"Good," said Apa. "You have done very well, my son." His brothers were happy to see him as well.

> *Then the great Father of all took the ball in his two big hands, breathed on it, and whispered his magical words. Then he rolled it around and around and around. Gradually it grew bigger and bigger. A great wind blew from Apa's lips, and the surface of the ball dried until the bit of mud that Bambooka, the little muskrat, brought from the bottom of the great sea, became the world on which we all live today.*

What lesson of life did you learn from the story? Why is this story important? Perhaps lessons such as "Do not give up on a problem," or "Try to focus on what you are trying to do while you're doing it," or "Try your hardest," or "Encourage others to try and offer positive encouragement," or "Do not be afraid to try to solve hard problems even if you fail the first time; try, try again," or lastly, "Know that you can face problems in life and seek the counsel of others." This little story is important as my story unfolds!

Some folks think holistic learning is a new concept, but for many Indian people this was, and is, not new. It is a way of life. For example, there are times for everything we do as Indian people. Some call this "Indian time"; others view Indian time as always being late, but I do not think so. If people are gathering food like berries, they are aware of the time of month and even the weather. They are also aware that man (the two legged) is not the only animal that likes to eat fresh yummy berries. Grizzly bear families (the four legged) like to eat berries as well, and for many of the same reasons as the two legged. The two families (two and four legged) collecting at the same time and at the same place can and does cause problems of dangerous interactions. To not know this concept of plant time, or not respect this knowledge, could lead to a bad outcome for both a human family and a grizzly family. We need a holistic knowledge of life. We need to pay attention and be respectful of Indian time: being at the right place at the right time.

CHAPTER 1

Traditional Knowledge and Education

I got my education from my culture. My teachers were my grandmothers and I am grateful for that.

—Mary One Spot, Blackfeet

WESTERN EDUCATION IN INDIAN COUNTRY

Euro-American education of Indian people was not so much about learning the three Rs as about saving the souls of Indian people. I am not sure that the first or present-day Indians needed or need their souls saved in the first place. In *History and Foundation of Indian Education*, Stan Juneau (2013, 5) writes:

> Indian tribes had their own education systems already in place prior to the landing of Columbus in 1492. Indian education consisted of specific roles played by each member of the tribe that centered on survival as a group of people. The transfer of knowledge from elders to the young, from men to boys, from women to girls, encompassing the history, culture and religion of each tribe, created an education curriculum that was passed on through oral tradition and practical, hands-on training.

In "The Indian Student Amid American Inconsistencies" (1978), Vine Deloria Jr. states that Indian education has been built on the premise that the Indian had a great deal to learn from the white man; the white man represented the highest level of achievement reached in the evolutionary process. The white man's religion was the best, his economics superior, his sense of justice the keenest, and his knowledge of history the greatest. The Indian's task was to consume bits and pieces of the white man's world in the expectation that someday he would become as smart. The totality of the white man's knowledge was supposed to encompass the wisdom of the ages, painfully accumulated by a series of brilliant (white) men.

What better way to "kill the Indian and save the man" than through "white" education?

Harvard College (1636) and Dartmouth College (1769) were established in part to provide white man's education to Indian men. However, it was through later boarding schools that culture and language for Indian people became key to assimilating children into mainstream culture. The boarding school experience for Indian children began in 1860 when the Bureau of Indian Affairs established the first Indian boarding school on the Yakima Indian Reservation in the state of Washington. The government BIA schools were established to prepare Indians for the mainstream workforce with European settlers. For example, girls learned housekeeping skills and to cook meals for white people. Boys also learned specific skills, like farming. No advanced learning or thinking.

Over time, because of the expense, the government reduced its influence on Indian education and paid religious orders to provide basic education in boarding schools and on reservations. Among the Western and Plains tribes, missionaries from all Christian faiths opened schools to "save souls." The government carved out reservations with various denominations of the Christian faith. Some reservations were given to the Catholics, some to the Moravians, and others to Protestant denominations. Fort Hall was awarded to the Episcopalians, who quickly established a boarding school for girls called the Fort Hall Episcopal Mission School (Just 2020). Eventually other schools were established, including the Fort Hall Boarding School.
The school was in operation at Fort Hall between 1880 and 1936 and upwards of 200 children attended annually. In 1937, the Lincoln Creek Day School operated only until 1944. It was part of a last-ditch effort from the Bureau of Indian Affairs to indoctrinate Shoshone and Bannock children into white culture. After this school, and a couple of others like it on the Reservation were closed, Indian children most often began enrolling in regular public schools (Just 2020).

It may have been, and perhaps still is today, that the missionary school staff needed their own souls to be saved. Missionary men and women came to lead a campaign to convert the Indian people to the Christian religion. Indian customs, beliefs, values, ceremonies, and education were ignored, discouraged, disrespected. The premise was that the Indian peoples' beliefs were inferior to those of the non-Indian. Some missionaries even believed that Indian spirituality was of the devil. Indian parents were given no voice in the educational process of their children.

The founder of Carlisle Indian Industrial School in Pennsylvania, Captain Richard H. Pratt, observed in 1882 that "Carlisle has always planted treason to the tribe and loyalty to the nation at large." The Carlisle philosophy was "kill the Indian to save the man (Slapin and Seale 1998, 338). I can never forget this very bad idea by the government of the United States of America. Led by a powerful nation, this "war" on defenseless children is still very hard for me to forgive. Nevertheless, forgive I must, if I am to move on.

Sadly, thousands of Indian children as young as three were forcibly ripped from their families. They were sent hundreds of miles away from their communities and reservations to be brainwashed by the boarding schools. Their hair was chopped off. Their medicine bags and beautiful beaded clothing were burned, and the children were forced to wear heavy woolen military-style uniforms even in summer. Uniforms, strict regulations, marching in formation, and all the trappings of the school were designed to create order and produce "character" so students would be equipped for a life in mainstream white America. At the boarding school, Indian children were forbidden to speak their Indian languages, forced to shed familiar clothing for white men's garments, and subjected to harsh discipline. A Lakota elder told me how when he was a young child, for days he was chained to a radiator in the basement for speaking his language.

Lone Wolf (Blackfeet) described his experiences in a boarding school: "Once there, our belongings were taken from us, even the little medicine bags our mothers had given to protect us from harm. Everything was placed in a heap, and set afire. Next was the long hair, the pride of all Indians. The boys, one by one, would break down and cry when they saw their braids thrown on the floor" (Slapin and Seale 1998, 338). Hair continues to be sacred for many traditional Indian people:

> Hair is considered sacred and significant to who we are as an individual, family, and community. In many tribes, it is believed

> that a person's long hair represents a strong cultural identity. This strong cultural identity promotes self-esteem, self-respect, a sense of belonging, and a healthy sense of pride. As part of the practice in self-respect, Indian children are taught to take good care of their hair through proper grooming. Hairstyles and ornamentation are guided by the values of our family and tribe. It is a form of creative self-expression that reinforces our connection to our family, tribe, and Creation. Some tribes will use two braids, while others will use three. Some families will paint their hair depending on the ceremony or their family's distinction. Women and men will adorn their hair with fur wraps, woolen wraps, feathers, fluffs, and bead work for war dancing and ceremonies. Braiding a child's hair is the beginning of establishing an intimate and nurturing relationship. An old grandfather first told about being forced to cut his hair when he was carted off to boarding school. Eventually, he told the story that his hair was cut in an effort to strip him of his culture and identity. Cutting his hair was their way of showing dominance over him through forced assimilation. He said that every time his hair was cut, he would cry, and every time he would cry, he would be physically punished. Unfortunately, being forced to cut our hair was a common practice in many institutions and schools across the country, and is still occurring as recent as 2018 (Sister Sky 2019).

They were denied the teaching of Indian elders, the company of kin, and the familiar foods, smells, and sights of home, and yet we wonder why some students ran away from school (Slapin and Seale 1998, 339).

The brainwashing of children still happens today. The federal government's annihilation of language and culture was enforced to break the hearts of Indian families. Boarding schools were also established in Canada and Australia by their federal governments for the same purpose. Children died, killed themselves, ran away, and were beaten badly for speaking their language. The newborns of young girls raped and impregnated by the clergy were thrown into fiery furnaces or killed in other ways and buried on the school grounds. Many of our grandmothers, aunties, grandfathers, and cousins were traumatized, beaten, and raped by both the male clergy and female sisters and other staff of the boarding schools (Cajete 1994; Dunbar-Ortiz 2014; Kirmayer and Valaskakis 2009).

In 2021, on the grounds of hundreds of boarding schools that were government funded and managed by religious organizations in Canada and America, thousands of unmarked graves of Indian children have been discovered. Their names and tribes are unknown. The world is just beginning to learn how many children have perished at the hands of religious managers in the confines of these boarding school. Our relatives experienced historical trauma, and that historical trauma has been passed down to succeeding generations, which is another tragic story.

The western education of Indian children marks a very dark and sad time in United States history. A disconnect with Indian values and education had begun. Following the policies and the orders of the official United States government, schools destabilized the culture of a whole generation of Indian people. At a great price, the Indian families and languages survived. The love and hearts of the people were strong and are strong even today.

In spite of the harshness of the boarding schools, some Indian people found success in learning. Children learned to write and read in English. They were exposed to new ideas and met new people, both Indian and non-Indian, and found ways to walk in both worlds, as one in the Indian world and non-Indian world. Not an easy thing to do then or now. I think of Mr. Jim Thorpe (Sauk and Fox), one of the most accomplished all-around athletes in history. In 1950 Jim was selected by American sportswriters and broadcasters as the greatest American athlete and greatest gridiron football player of the first half of the twentieth century (Slapin and Seale 1998).

> The discovery of Thorpe at the Carlisle Indian Industrial School in Pennsylvania, the government-run boarding institution for Indians he attended from 1904 to 1913, between bouts of truancy, is a well-worn story. In 1907 he was ambling across the campus when he saw some upperclassmen practicing the high jump. He was 5-foot-8, and the bar was set at 5-9. Thorpe

asked if he could try—and jumped it in overalls and a hickory work shirt. The next morning Carlisle's polymath of a football and track coach, Glenn "Pop" Warner, summoned Thorpe.

Thorpe was number one in four Olympic events in 1912 and placed in the top ten in two more—a feat no modern athlete has accomplished, not even the sprinter and long-jumper Carl Lewis, who won nine Olympic gold medals between 1984 and 1996.

[Thorpe was born on] the Oklahoma frontier, orphaned as a teenager and raised as a ward of government schools . . . [he was] uncomfortable in the public eye. When King Gustaf V of Sweden placed two gold medals around Thorpe's neck for winning the Olympic pentathlon and decathlon and pronounced him the greatest athlete in the world, he famously muttered, "Thanks," and ducked more illustrious social invitations to celebrate at a succession of hotel bars. "I didn't wish to be gazed upon as a curiosity," he said (Jenkins 2012).

CONTEMPORARY INDIAN EDUCATION AND THE ROLE OF GRANDMOTHERS

The non-Indian people say that God made the earth and the universe, but we Indian people know that Coyote made it first.

—Oral history creation stories from Navajo, Ute, Salish

My story, which continues with the Sho-Ban students on the Sho-Ban Reservation, is not filled with education theories. However, if you look thoroughly, these concepts are hidden in this story. Much like the Shoshone-Bannock story that I shared with you at the beginning of this narrative, you have to think about it and find your own pathway to teaching/learning. This is a story about the love and beauty of Indian students on a wonderful adventure of learning. My students asked important questions and made their way in a complex world of space. More importantly, they learned how to meet an impossible deadline with NASA and accomplished the mission that was required of them. Indian students can and will do great things if they are *believed in* and are *shown* that there are indeed different ways of knowing. Their grandmothers are the important link to classroom learning and home learning.

An Indian grandmother is respected, and at times feared when made angry, and she is greatly loved. Grandfathers are respected as well. But clearly, to me, grandmothers are in a different category. The Indian grandmothers I know love and care about family members in a way I have not seen outside the extended family of a reservation. If the family has a problem, grandmothers know how to fix it.

I met Lilly, the grandmother of one of my students. Lilly grew up in a transition time. It was a time when the Indian people were still holding the knowledge of the old ways and were learning what the new ways were about. Lilly went to a boarding school in Riverside, California. She was not allowed to speak her Indian language or dress the way she preferred. Her braids were cut. Her education was focused on changing Indian children into white people. No clear academic choice was given. Girls learned homemaking skills and boys were taught farming skills.

Lilly survived the Bureau of Indian Affairs boarding schools and to this day has a curious sharp mind. Lilly is short and round and brown and has a wide friendly smile and gentle laughter. Her gray and white hair is pulled back in two braids. Lilly will tell you that she has earned all the lines on her face. Lilly solves problems and is loved by all. With a very limited income and resources, she knows how to help her grandchildren get what they need. She helps them move forward in their lives. Lilly is super smart. She is great at asking questions about her family, and she takes no unkind attitudes or words about them from anyone! She argues with countless folks when she sees injustice being done. She has earned the respect of scores of Indians and non-Indians.

The BIA and related Bureau of Indian Education are government departments that deal with Indian funding and policies, including education. Indian people say that BIA stands for "bossing Indians around." I would advise anyone not to "boss" Lilly around. Lilly is a great force for education for not only her kids and her grandkids, but for the community as well. I seek advice and counsel from Lilly all the time. It is a great honor for me to listen and learn from this beloved Elder.

Education continues to be important to Indian people. Without the ability to reason, question, share data, adapt, and invent new tools, Indian people never would have survived. Indian people not only survived, but they thrived! Indian people lived and, in some cases, still do live, in some of the most seemingly inhospitable places on Earth. For example, the hostile far northern latitudes of the Arctic, or the barren deserts of the Southwest, or lands surrounded by sea.

They learned to live in partnership with their environment. Using the stars as maps, they were expert navigators. Local medicines and medicine plants cured hundreds of illnesses. Some Indian healers actually discovered how to release pressure on the brain through trepanning. Applying their scientific principles, they learned to preserve food, tan hides for clothing, and use local construction materials to build shelter. Theirs was a curriculum of surviving in their environment, and application of that knowledge meant life or death. With songs and dances to honor a Creator or the great mystery of life, they learned to care for one another. They learned until they got it right. Many non-Indians still call some of these places "wilderness"; however, I call it our true paradise and home.

One of the most powerful forces in the universe is an Indian grandmother. I learned from my grandmother that there was more than one way to know and learn. As I went on in life I learned about other ways of knowing. I was invited to Canada for a First Nation Conference on education. During the conference I attended a lecture by a Mi'kmaq Elder named Albert Marshall. The elder had a PhD in life and talked about a concept he called "two- eyed seeing," which he explained as "learning to see from one eye with the strengths of Indigenous knowledges and ways of knowing, and from the other eye with the strengths of Western knowledges and ways of knowing . . . and learning to use both these eyes together, for the benefit of all" (Marshall 2020).

This made sense to me, as I thought about it. This was how I was learning, and I asked the elder two questions: "Can I not only learn this way, but can I teach this way?" and "Can I share your good words with others?" He took his time to look at me and said, "Yes, you can share." And he asked me why wasn't I already teaching students this way. I told him that I would share his words and start to teach this way . . . now! He was pleased with my answer. I owe a lot to this elder. We, as Indian people, owe a lot to all of our Indian elders.

The need for a culturally relevant education is as strong today as it was in the late 1800s.

Michelle Whittingham (Cherokee) wrote, "I want to help ensure that Indians don't forget who they are, where they've been, where they're going. I want to encourage them to keep moving forward" (Gilliland 1999, 1). Hap Gilliland also states, "Education is like a choir, and a choir is not so good if all members sing alike" (1999, pp.1–2). He describes education as the blending of many different voices that produce beautiful music. Today, educators call this concept Culturally Responsive Teaching. I apply this Culturally Responsive Teaching (CRT) in my classroom with my students. We make meaningful, beautiful music and connections with what the students learn in the classroom, their community, culture, and life experiences (Harris 1995).

The purpose of education is, first, to communicate to new generations the knowledge that people have amassed through the ages and, second, to provide them with the ever changing and increasing body of knowledge necessary for them to adapt to a changing world. Gilliland writes, "All students must begin by understanding their culture, and be confident in applying the knowledge, and cultural understanding of their own people, to many circumstances. To bring about harmony, the teacher must understand the culture and values of every student" (Gilliland 1999, 1–2). Gilliland explains that the majority of teachers of Indian students do not know the values or cultures of their students, or how their instruction can be adapted to the students' needs and ways of learning. Many of the schools discredit Indian cultural heritage, damaging the students' self-esteem and motivation.

Education today, for some Indian people, is a long, hard struggle. Many Indian students struggle with mathematics, science, or history exams. For some young Indian scholars, the struggle is not so much the subject material as it is having a positive attitude toward who they are. The struggle is also sometimes one of low teacher expectations. This is one of the many areas where Indian grandmothers and grandfathers shine. They love their grandchildren for who they are, beautiful people. Grandparents know a young person's potential for good, and as grandparents, have high expectations for them. They have been around the sun enough times to see patterns of good, compassion, and also, sometimes, patterns of despair. They know potential when they see it. They know the struggles the youth face, and they believe in the goodness and potential of their grandchildren. Sometimes, I wonder if teachers could believe in their children and students as much as grandparents do.

In our country, supposedly the richest country on earth, Indian dropout rates are the highest of any ethnic group (National Center for Education Statistics). A 2015 *US News and World Report* article reported that "67 percent of Indian students graduated from high school compared to the national average of 80 percent."

Researchers have found that it has been a common assumption that Indian students who were dropping out were failing academically. In 1997, Deyhle and Swisher reported that the academic achievement of dropouts did not differ significantly from those who remained in school. They found that 45 percent of the dropout students had a B or better grade point average. *The Senate Special Committee on School Performance* (1989) asked high school seniors in Alaska why their peers dropped out of school. They consistently reported unsupportive teachers, inability to memorize information required to pass a class, and boredom. Not lack of gray matter power!

I respect teachers, and I am a teacher. I know how hard many of us work. Most care deeply about our students and work incredibly hard to help all students. However, a teacher's attitudes toward Indian students are critical to a student's success in school. As in any profession, not all teachers have a positive attitude toward their students. A positive attitude from a teacher is important for any student, but it is critical for the Indian student.

A well-respected Indian teacher and researcher, Jon Reyhner, wrote,

> A teacher's attitude is . . . contagious. . . . A teacher who can earn the respect of Indian students and who can show them they are respected for what they are is well on the road to giving those children success in school. . . . Too many teachers and other well-intentioned individuals look at the physical surrounding

> in which Indian students live, the prejudice they face, their problems in school, and they sympathize. They feel sorry for them. These students do not need sympathy; they need something to be proud of. Pity and pride do not go together. (Reyhner 1994, 77–83)

Our NASA story is about this pride: pride in one's tribe, family, and self; pride demonstrated in the eyes and smiles of parents and grandparents. It is the pride and a belief in young Indian students. With help from Indian grandmothers like Lilly, I am convinced that with pride and respect, the young Indian students in my class could do more than just pass the standard test of knowledge given by the BIA. I was soon to discover that the pride and respect I have for my students was soon going to be put to a great test . . . by NASA.

CHAPTER 2

Teaching Science at Sho-Ban

Native science evolved in relationship to places and is . . . instilled in a "sense for place." Therefore the first frame of reference for a Native science curriculum must be the "place of the community, its environment, its history and people."

—Gregory Cajete (1999, 47)

I completed my teaching requirements and applied for a position as a science teacher at Sho-Ban High School on the Sho-Ban Reservation. When they offered me the position, I accepted. My soul was happy, as I knew this is what I needed and wanted to do with my life. My students were, and are, smart. I was required to teach Idaho's state science objectives. The majority of the students learned them quite easily. I had an open classroom and word got out throughout the reservation that I made coffee in my classroom, and I always had a pot bubbling. Coffee is important to residents of reservation communities, and offering a cup of coffee to visitors shows a caring nature. The Salish people of Montana even incorporated coffee into one of their dances. As a result of the coffee, I had many visitors. One such visitor was a cultural teacher on the reservation. His name was Lavern Broncho. Lavern and I became good friends, and he taught me many things about his reservation and reservation life. One of things he taught me is that Sho-Ban Indian members consider salmon in high regard. This magnificent fish holds an essential cultural and nutritional position. He gave me an important question to think about: How could I inspire the young students at the Sho-Ban school to learn and to help the young salmon grow and survive? I thought about this question for a long time, as both students and salmon were and are still important to me.

In his book *Look to the Mountain: An Ecology of Indian Education,* Cajete (1994) suggests that Indian students are inspired about education when the content makes sense to them. Attention should be paid to the practical needs of the Indian community and learning to be a productive member of the Indian community. Knowing Cajete and his work regarding this aspect of education, I decided to do two after-school projects in the same time frame: 1) NASA for students who like space, computer work, and working more inside at our lab; and 2) a project for students who enjoy working with their hands and camping in the mountains in summertime. Not all high school–age Indian students like to sleep outside and fight mosquitoes!

I asked the Indian fisheries department to help me answer Lavern's salmon question. The fisheries department was pondering similar questions about helping salmon. They shared an idea of growing salmon in containers and then releasing them. Their model was to use old repurposed refrigerators as growing chambers. Fisheries staff had recently participated in a workshop in Rock Springs, Wyoming, and were shown how a veterinarian was successfully applying this idea to brown trout. I called the veterinarian and he agreed to help our school remake old refrigerator boxes to use for salmon. Because salmon are an endangered species, we had the tribe's fishery department help us access salmon eggs, including all required permits. The students were highly successful making salmon boxes out of old refrigerators. I was still teaching the state requirements, but in a way that was interesting and relevant for my students.

The top lid of the refrigerator box has sixty 3.5 x 13 millimeter slots for water circulation and a swim-up fry escape passage. The refrigerator also restricts predators and works as a desilting mechanism. The flap of the top lid opens into the incubator, which holds sixty-nine 2 x 2 millimeter vents for circulation, ventilation, predator protection, and silt retention. The top compartment can hold one or two layers of approximately 250 salmon eggs or 500 trout eggs. Typically, hatch success in the boxes averages from 75 to 95 percent. Fry that successfully leave the box and enter the stream average from 20 to 50 percent of the original number of hatchery eggs. The streamside incubator boxes provided a secure environment. The students modified the interior of the refrigerator for the incubator. Acrylic dividers and rocks were placed in the bottom of the refrigerator so water flowing through it created currents similar to those in a small stream. Water from the stream was supplied using 1-inch-diameter polyvinyl chloride (PCV) piping. A 2-inch-diameter outlet pipe was used to

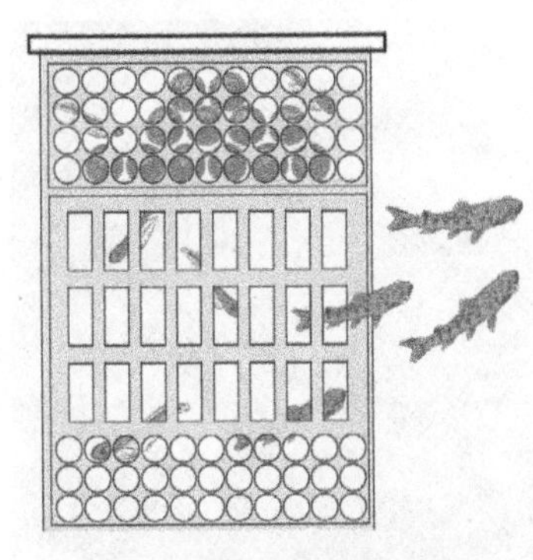

Whitlock-Vibert Box, side view (*left*) and front view (*below*). Courtesy Idaho National Laboratory.

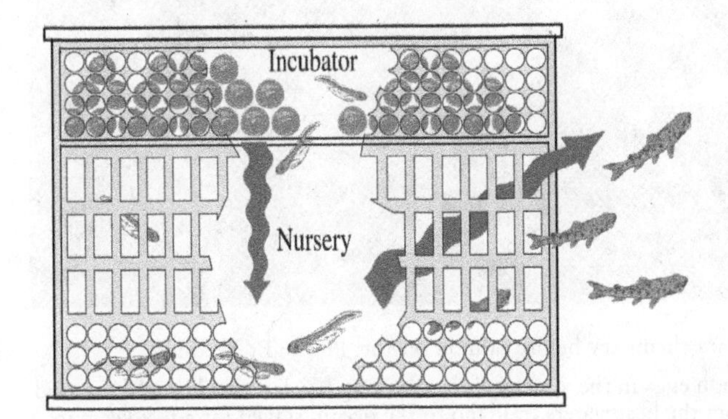

Placing the Whitlock Box streamside. Photo: Ed Galindo.

(*Above*) Testing the stream chemistry before salmon release. Photo: Ed Galindo.

(*Below*) Placing the salmon eggs in the stream. Fish, as young fry, left this box and traveled over 900 water miles from the headwaters in Idaho to the ocean, stayed for a few years in the ocean, grew huge, and returned to the boxes in Idaho. Photo: Ed Galindo.

control the water level and allow the fry to exit the incubator. The total cost of the converted incubator was thirty dollars. To prevent it from rusting, the box was placed at the side of the stream on a pallet. Bushes and foliage camouflaged the incubator box.

Anywhere from twenty to forty thousand eggs can be grown in the incubator using a little box called a Vibert box. The boxes fit along the sides inside the refrigerator-incubator. Unlike hatcheries, the incubator-refrigerator allows the eggs to survive in an almost natural environment. Once the eggs are placed in the box, they are not handled again by humans. Much like natural spawning, eggs in the boxes are subject to random mortality, which allows the stronger, smarter fish to develop greater survival skills. The new fry protected in the incubator develop a more advanced yolk sac, producing stronger mature fry that, after leaving the box, have a better chance of survival from predation or natural causes.

In this program the students were learning to raise salmon. Salmon, an excellent source of protein, are a cultural animal for the Sho-Ban. When I brainstormed with Indian members about student projects, we decided that this would be an excellent student project, with young Indian students helping raise and protect young salmon. As well, the Elders wanted to ensure that students could maintain the resource for the future. It was a beautiful project for all involved. The students loved this project as much as the NASA project. Both projects forced the students to think of things besides themselves, which is an important concept for learning and teaching. They loved it also because they were raising a spiritual and cultural resource that their people had depended on from ancient times. Some of those same students went on to university to become fish biologists and work for Indian fisheries.

What I learned while teaching on the reservation was to ask the community what was important to them, and then ask and listen to what the students wanted to do, then look for a compromise in my teaching time and subject matter. Learning about salmon and NASA just sounded like fun, exciting challenges.

As I reflect back on my life, I think one of the biggest influences on how I might best teach young people was the lesson wild horses from the Owyhee desert taught me while I worked as a cowboy on a ranch in southern Idaho. The ranch had thirteen hundred head of cows and employed over eight full-time ranch hands or cowboys. The ranch also had five- to six-hundred head of younger stock that they fed out and then sold in a feedlot on the ranch. When I was fifteen years old, I was employed as a full-time ranch hand. Every day ranch hands did all kinds of demanding chores that a ranch required. At fifteen, I believed that if I worked hard, I could earn the respect of the fellow ranch hands. I knew how to ride horses well. That summer, the ranch management decided that the ranch needed more young horses. We had twelve horses, but they were getting old, so we needed some replacements. The management learned that the Bureau of Land Management (BLM) was giving away horses. They were young, very wild horses captured off Idaho desert lands. The horses were three- to four-year-old animals that had never seen or interacted with humans.

The ranch acquired fifteen horses, and it was determined by the other ranch hands (older men) that another young boy and I should break the horses for the ranch. "Breaking a horse" means to train it to accept a saddle and rider. They were fully grown and very strong horses. Since they had never before seen humans, or had human riders on their backs, they were frightened of people.

I inspected the horses in the corral. They were extremely quick and strong. On the first day of my horse training job, the other ranch hands and I were going to rope one and place a saddle on it. It was my job to ride the wild animal until it stopped bucking. I was thinking to myself, "This is not so bad" . . . was I wrong! I mounted my first horse, and the first thing it did was use its back legs to kick me off! The horse was not finished with me. Once it had me on the ground, it chased me like a dog would and tried to bite me. The chase continued until I rolled under the fence and out of the corral. This happened over and over again with each horse.

My first day breaking wild horses ended with me being broken. I learned very quickly that wild horses were faster than me, stronger than me, and knew more than I did about survival and life. I learned that I had to gain their respect to train them to do what I wanted. This happened in a variety of ways. The best method for me was to build a slow, honest relationship. Getting to know the horse, and slowly letting the horse know me, caused far less drama.

I never forgot this idea, and on my first day on the reservation working with the students, I remembered the wonderful lesson that these wild horses gave me. I had to build a trusting relationship in order to survive teaching. I did this with the students as I did with the horses. It worked. Both horses and students would do anything for me if I was honest with them. I learned for both groups about how best to teach. The students I worked with were stronger, faster, and wanted me to respect them, much like the wild horses. Once trust was built, we could begin learning from one another.

BUILDING TRUST WITH NEWTON'S LAWS OF MOTION

Doing a physics experiment in my classroom was another way of teaching/learning and building trust. The experiment examined a complex problem using simple things found in the kitchen. Students liked to do this project. I have found that this is an excellent example of putting the fun back into teaching a scientific concept. This experiment is a cool physics demo on why you need to wear seat belts, as well as some other life lessons. The materials needed include an old sheet, two raw eggs, a four-inch Ace stretch bandage, three sets of safety glasses, and three students to volunteer at the front of the classroom.

This experiment also demonstrates why it is not safe to drink and drive. There are certain laws we need to obey. The laws of physics are not an exception. There are alcoholism problems on the reservation, and many students have heard the pitch of not drinking and driving. At this age, most teenagers know everything in the world—well, they think they do. So, this demonstration weaves a couple of stories: the dangers of drinking and driving and the importance of wearing seat belts. For every action there is a reaction, and momentum.

Isaac Newton, a seventeenth-century scientist, came up with some ideas about why objects move (or don't move) as they do. Newton's first law is called the law of inertia: "An object at rest tends to stay at rest and an object in motion tends to stay in motion with the same speed and the same direction unless acted upon by an unbalanced force."

Okay, you say, so what? Well, let's say you are in an automobile while it is braking for a sudden stop. All objects in the car tend to keep moving forward unless acted upon by an unbalanced force. The unbalanced force in this case is defined as braking the wheels to change the car's state of motion until it stops. However, there is no unbalanced force to change a passenger's state of motion. What will stop the passengers' motion if you are riding in a fast-moving car? If one is wearing a seat belt, it will help stop your forward motion. If passengers in the car are not wearing a seat belt, the window, car body, or your head will! So, as you can see, seat belts can be used to provide safety for passengers whose motion is obeying Newton's law of inertia.

After this explanation, I have two students, wearing safety glasses, come to the front of the classroom. I ask each student to hold one end of a sheet a bit higher. This makes a small cup in the sheet. The third student, also with safety glasses on, has an egg in her hand that represents a loved one in a car with no seat belt on. I ask her to throw the egg as hard as she can at the sheet the other two students are awkwardly holding. The beloved egg hits the sheet and rolls down into the cup part of the sheet. Everyone, including me, is relieved that the egg does not break. The sheet is, of course, symbolic of a car's airbag inflating. We have demonstrated Newton's first law. Then we do it again, but this time I cover one of the egg thrower's eyes with four-inch-wide stretch gauze. Why? I tell the class that this will demonstrate lack of full human ability to a reaction, like when one is texting and driving, not wearing a seat belt, or driving after drinking alcohol. I spin her around slowly to further add to her disorientation.

Now we try it again. I tell her to quickly throw the egg at the sheet. Most of the time, the egg misses the sheet and hits the wall or the chalkboard behind the sheet, or one of the students holding up the sheet. This, of course, makes a mess of the egg and elicits a great deal of laughter from the rest of the class. Once in a while the egg hits the sheet just fine. When this happens, I can talk about luck, the luck of not getting hurt when one is driving and drinking. This happens in life. This experiment is a way of talking to students about the hazards of drunk or distracted driving without so much preaching. We also discuss the reason why we wear safety goggles while doing experiments.

In summary, the egg experiment illustrates to the students the many different ways of learning and teaching, and that learning can be fun and takes many forms. There are many lessons to learn here. One is to be careful of our "eggheads," as they can crack very easily.

CHAPTER 3
Mars

Getting to the sky is no problem. Come, I will show you how.

—Coyote to his wolf brothers, Wasco Legend

How did a small reservation school in Idaho get a scientific experiment on a space shuttle? And what in the world would enable Indian students and a reservation teacher to be players on an International Space Station mission with NASA? The story began in 1985, when Dr. Ali Siahpush repeatedly left me phone messages. He wanted to meet with me and my students. I thought he was a science textbook salesman. I had purchased the textbooks for the year and had no resources for more.

But Al was extremely persistent, and one spring day, I finally called him back. Al excitedly explained that he was employed by a government nuclear research lab as an engineer. He was also a member of a group of NASA folks called the Rocky Mountain Space Consortia (RMSC; now the Utah NASA Space Grant Consortium), which was made up of several universities in Utah, Idaho, and Colorado. The consortia has many missions, including education and outreach. RMSC was concerned that not enough students, particularly minority students, were entering science and engineering fields. If this trend continued, it could lead to a drying up of the science and engineering professionals for our nation. NASA and RMSC were education partners. The concept here was that if dollars could be spent early on to motivate students, they might be inspired to attend college, study engineering and science, and thus be eligible to work for NASA in the space science field. This would help fill the education pool. They wanted minority participation, and Al wanted to know if our Sho-Ban school might be interested.

Al had become the spokesperson for NASA and the RMSC group in Idaho. We decided to meet and chat. Al was from Iran, and now he was an American citizen. Al had a bushy black beard and hairy arms. His hair was in a ponytail. My students named him "the hairy Indian." (Among some American Indians, this is a sign of acceptance to the group.) Al liked his name and smiled.

Over the years Al and I developed a friendship, and my students and I attended all kinds of NASA workshops, mainly with the Utah NASA Space Grant Consortium. Then one day we had a new conversation. What Al told me made me laugh . . . his group wanted my students and me to go to *Mars*! Al wanted me to recruit students to go to a Mars workshop in the spring of 1993 to be held at Utah State University. The university is about two or three hours away from the reservation by car. We would be guests of the consortium and NASA. I was excited, as this was something new to offer my students. It sounded fun, and it was free.

THE PITCH

I made a pitch to my junior and senior science class the following day. This was a class of bright, young chemistry students. I began by asking how many of them would like to go to Mars. Not one hand went up and some just stared at the ground. I asked my students if they knew where Mars was, and a few volunteered that it was a planet in space. Some knew it was our sister planet. All knew it was far away. Some dedicated students just wanted to work on their homework assignment. Such dedication!

The next day, I asked again, how many would like to go to Mars. Again, no hands went up. Silence filled the chemistry room. Finally, one hand went up from a student named Sarah. She asked, "It's a long way away, right?" I said yes. "Then why don't you go and stay awhile!" This brought on much class laughter, even from me. Sarah was a beautiful young lady with jet black hair that was cut short. Sarah had, and still has, a wicked sense of humor.

I have a bit of a weird science reputation of asking my students to do things that will shock them out of their comfort zone. Why? To get them to think and do things! For example, before a science demonstration, I ask the students specific questions as a focus point for them to guess (hypothesize), and then show them what could happen. All my science demos are relatively safe—like my famous egg and sheet demonstration. How does this relate to our Mars workshop? Well, my students know me well, and they knew there was more to this story than just going to Mars. They wanted to know, "Why would I want them to go?" I liked this question a lot. Why do we ask students to do anything?

I told them that my new friend Al, "the hairy Indian," had invited us to go to a Mars workshop out of state. Some of the students were fine with this. Others questioned using their summer break for a school-related project. Many were nervous about going out of state. "What will we do once we are there?" some of the students asked. I told them that I was not sure, but I knew it would be fun. Some of the students asked if shopping could be done after the workshop. I said, "You bet." Reservations are mainly rural communities, so going to a town four or five hours away is a big deal for a student. And since this was before online ordering became so common, shopping was a big draw. Four brave Sho-Ban science students volunteered to go to Mars and do some shopping.

MARS WORKSHOP

The folks from NASA and Utah State treated us extremely well. The workshop was fascinating to me as a science instructor. We learned about Mars and we asked many questions: What happened to the water the planet seemed to have? Did it have life? Could our planet "lose" its water? The students learned that Mars has the largest volcano in our solar system, very large canyons—again the hint of water being on the surface at one time—and a mysterious "face" on its surface. My students were hooked now—a face on Mars, no way! Who built it and why? They learned that the face was made by the forces on the planet, the force of wind and perhaps water. The force of erosion made the face, and the face was formed by our imagination the way we see faces in rocks on Earth.

My students and I were part of the discussion of Mars exploration. We heard debates by NASA officials and the scientific research community about manned missions versus unmanned robotic missions. My students asked questions, and I was proud to see they were engaged in debate with people from all over the world. My students asked, "Why were we going in the first place?" They also asked about the dangers of bringing disease to the planet of Mars from our world. Indians know the contamination story well because this happened to many tribes with European contact. Viruses and bacteria can be deadly to life forms that have little or no resistance. My students' voices were heard that day at the Mars NASA workshop. The answers varied. Some researchers had similar concerns. Some had not thought of viruses or bacteria as much of a problem. Others wondered about transporting microbes from Mars to our home.

When the conference was over and our shopping was complete, we headed back home to the reservation. My students were excited and we talked about what we had heard and seen. One of my students, Jake, commented that he liked how the NASA people respected his opinion. Jake was a huge young man, well over six foot five and over 240 pounds. He was an excellent student with an inquiring mind, and he was a meticulous worker. His eyes never missed the finer details of his surroundings. Although he wore his hair shorter than the other young men, he was truly a traditional Indian member. Sarah, with her shy, quiet ways and soft voice, was not one to jump into any discussions unless asked. She added that she was happy that the NASA people asked her what she thought.

We met people from all over the world. We heard many new ideas, and they were exciting to hear. We also noticed that at this conference there were no other Indians, and very little minority representation at the discussion tables. Jim, a tall, thin, quiet student who is built like a champion wrestler, with large arm muscles and strong hands, wondered why we talked so much about all kinds of things. I said, "To learn." Jan, with long black hair and a wonderful smile, raised by her grandmother in a traditional manner, asked why there were few female scientists present. She shared that her grandmother told stories about Indian people who have been explorers, scientists, and teachers living off the land since time immemorial. We all discussed survival techniques that our families and relatives practiced, like tanning deer hide or gathering certain plants for healing. We reviewed how to get what we need to live respectful lives, and be thankful for what we have. We debated the idea of using just what we needed, never taking too much, and always giving thanks. Perhaps we could do an experiment in our science class that would show this example to our new NASA friends. We were thinking, "Could astronauts live or survive off the 'land' as they were traveling through space?" This is an old concept for Indian folks, but new to the "final frontier space folks" we had just met.

We talked about the idea of just exploring space, not building a fort or planting a flag and claiming the entire planet for just themselves. "What a very silly concept," we all said. We agreed that traveling through the universe is fine, but staying and taking over a planet is another issue. We discussed the hideous history of Indian peoples' experience with this idea, and the fact that Indian people did not fare so well.

CHAPTER 4
KC-135A

> *Questions about our origins were once regarded as the territory of philosophers and theologians; but gradually the answers have been provided by science; speculations have been replaced by hard facts.*
>
> —Stephen Hawking (2017, 199)

Al called me again after we made it home. The NASA and Rocky Mountain Space Consortium folks were pleased that my students and I had come. They were very impressed with how our students conducted themselves in the workshop. Would we be interested in more space-science-student opportunities?

I said, "Sure, what did you have in mind?" Al told me of a rigorous competition that NASA was about to embark on. It was called Reduced Gravity Student Flight Opportunities Program (RGSFOP). I learned that NASA, like other government agencies, loves acronyms. RGSFOP was but one of thousands of acronyms for programs that NASA supports. A successful experiment would give students the opportunity to fly in space on one of the space shuttles.

RGSFOP wanted our students to design an experiment that would be carried out on a NASA aircraft, the KC-135A. Based in Houston, Texas, the KC-135A is a special aircraft. Its main mission is to train astronauts, to test science and engineering designs, and to conduct experiments in microgravity conditions. The KC-135A simulates space flight by flying a series of parabolas, which are like the loops on a roller coaster, but on the edge of space. The KC-135A flies upward at a steep 50-degree angle over the Gulf of Mexico to about 35,000 feet (10.67 kilometers) and then down at a very steep angle to about 15,000 feet (4.57 kilometers). Those aboard experience weightlessness in the peaks and valleys of the parabolas. Researchers and astronauts test experiments and themselves on the flight before they fly in space. NASA describes the Boeing KC-135 Stratotanker on its website:

> Besides being used extensively in its primary role as an inflight aircraft refueler, [it] has assisted in several projects at the NASA Dryden Flight Research Center, Edwards, California. The NASA Reduced Gravity Program began in 1959 and the KC-135 was the perfect aircraft. The Boeing four-engine turbojet Stratotanker was originally designed for in-flight refueling, and later as a 707 for commercial flights. Further modified to meet NASA's needs, the KC-135 is used to understand the role of gravity on humans and hardware in space (NASA.gov).

The KC-135A has another name, given by the astronauts that train with it—the *Vomit Comet*.

The KC-135A student program started in 1995. Prior to this time, only NASA astronauts or community personnel could fly on this magnificent bird. The primary goal of the new program was to involve students in research areas of interest to NASA while providing an invaluable experience. Students would learn not only about space, but also about the NASA family as well. To be selected to fly aboard the KC-135A, student teams must initiate an idea for an experiment, and then prepare a very detailed proposal for peer review by NASA's scientific and engineering teams. The proposal is evaluated on scientific merit, design feasibility, fabrication, and compliance with NASA's strict flight safety experimental protocols. The ultimate goal for students is to conduct real research by applying science and engineering principles to their projects. They learn as much as they can while in a microgravity state, and then share the results with others. This is not a NASA mission for student joyriding, but it is quite fun. It is a rigorous program. Only the best of the best science teams in the nation would be selected.

Al was looking for minority participation and found it at our local reservation school near his lab. Although there were plenty of minority students in the area, outsiders are not always welcome in Indian communities, frequently for good reason, so he ended up working only with Sho-Ban students.

Although Al told me about the program, he did not suggest we apply, since only college teams were eligible to apply. I picked up an application from a NASA information person. I believed my students—mostly seniors preparing for college—could come up with a viable experiment to perform in space. High teacher expectations are important; I believe you need to show students that you believe in them. I told Al that we were going to apply as Sho-Ban High School and I thought we could win a spot with NASA. He liked the idea very much, but thought we may have trouble with the NASA protocols, as we were a high school team.

THE PROPOSAL

My students and I had to come up with an experiment that would be of interest to the NASA community. I thought about not only the NASA community, but the Sho-Ban Nation as well. We would then engage in open competition with schools across the nation. Not high schools, but universities. Al asked if we could do it. I said, "Sure, I have been thinking about an idea." We would give it a try.

As I dreamed about space and space travel, I thought about several other things. We were a reservation high school. Could we compete at a collegiate level with science and engineering students? I knew we could. I believed in my students 100 percent. As I was teaching and learning science on the reservation with my students, I concluded many things: My students are very smart. They know a lot about life. Life on the reservation is beautiful but can be challenging at the same time. My students know how to survive in all kinds of environments. I am not talking about outside environments; I am talking about environments where poverty and at times violence exists, environments where money and the resources that money can bring is in very short supply. I knew this because I had come from the same difficult environment as my students. I know that they can find humor and beauty in all this chaos of their lives.

There are some huge agriculture operations in the area of Idaho where the Sho-Ban Reservation is located. To force the Earth to produce more than she needs or wants to, farmers use phosphate fertilizer on the crops. There is a phosphate mine on the reservation, which is leased to a private off-reservation company that makes fertilizer and related products around the world. Many of my students' parents or caregivers worked for the mine. Would my students be more interested in phosphate chemistry, since this is a subject that they could see relevance with in their lives? Would chemistry and STEM subjects become more real if the students could see real-life applications on the reservation? As a school, we built old refrigerator salmon-rearing boxes, using what was on the reservation and working with the reservation community. It is hands on, brains on learning.

I saw a teaching opportunity developing with our space experiment. Several impending questions that pertained to Mars were lurking in my mind. If folks are going to travel to distant planets like Mars and beyond, what are they going to eat? If they had access to water, could they grow food? I knew that astronauts could carry processed food, but on an extended mission of three to four years, people would need to know how to live off the land, in-situ. Indian people have been doing this for centuries. The idea was not new; however, the space environment was new.

I had a series of debates and discussions with my students before we decided on a plan. The students and I talked about why people want to go to space to begin with. Initially and early on in the discussions, many students didn't understand why humans would want to take over another planet. They believed, like I did, that we should *take care* of our home world and then visit others, but not to take them over. And they came up with many good reasons. For example: Humans like to explore. Humans like a challenge. Humans like to have "fun floating in space." Humans like to "claim" land that does not belong to them. (Indians had experience with this idea, and it did not turn out so great for Indian folks). Tisha, a real bookworm with an unbridled imagination, thought it would "be awesome" to live on Mars. She told the class that she read someplace that if earthlings repopulate Mars, it may save the species from extinction if Earth was decimated by climate change or nuclear war. She may have been influenced by science fiction novels about Mars.

I knew that for all our technology and knowledge, we are still dependent as a human species on about four inches of good topsoil and rain; without both of these we die, as we need to eat and drink fresh water. I shared this thought with my students and we had several discussions on this topic. We debated living in space, and how best to do that, and why? The reason that came up the most was that humans who traveled great distances in space needed to be able to live off the land. They could not take all the food they needed for a five-to-ten-year mission. How to "make a living" in space emerged as our class focus. This is what we wanted to understand and do with our NASA experiment.

Finally, we decided on a research question: Could we make liquid fertilizer in space? And if so, could this space fertilizer be used to grow food? The students and I knew that the reservation's phosphorus mine produced fertilizer for agriculture. I queried the students: Could we grow food in space, using fertilizer that we made in space? The students liked the idea, and it made sense to them. I reviewed videos of how water reacts in a microgravity environment; the water forms a small round bead. Could we get this version of water to mix with fertilizer and have a useful product to help grow food in space?

We planned to submit to NASA an experiment in which we would mix phosphate ore, finely ground up like flour, with water. We thought that when men and women travel across the universe, the ability to mix fertilizer and water in space would be useful in growing food. My students could see farming being done on the rez every day. The farmers used what was at hand, and we used what was at hand to learn.

SHO-BAN TECH

What chance did we, as a reservation high school, have in an open science competition with college students? I thought we had a *great* chance. At this time, a colleague of mine at another school told me, "You Indians only get things because you are Indians." When I heard this statement, my blood boiled. I did not agree with this racist statement at all. This statement reflects the stereotypes that many people have concerning Indian people. It is erroneous, wrong, and unethical. He knew about the competition proposed by NASA. He said that he'd bet one dollar that we could not win a ride on this bird if NASA didn't know we were Indians. I took him up on the bet.

When we applied for a flight on the KC-135, I did not mark our ethnicity, or that we were a high school. There was no such box to mark. When asked the name of the college or university, I called us

Sho-Ban Tech. I thought that name was cool! Much cooler than our actual name, Shoshone-Bannock Senior/Junior High School.

The deadline to apply was fast approaching. My students and I met daily and continued to discuss our ideas. The students helped me brainstorm ideas for NASA. Some were skeptical. For example, Sammy wanted to know, "Why would anyone want to fly in a microgravity environment anyway?" Joelle responded, "We need to test out our hypothesis, to see if our project would work in space before it was supposed to fly on the shuttle." Anthony replied, "Any good science project needs the hypothesis tested in lots of ways." All the students thought this would be so much fun, and it was! I completed the application and mailed it to Johnson Space Center (JSC) in Houston. We had submitted a research project to one of the most remarkable institutions on our planet. We could hardly wait to see the results! In anticipation, every day the students gathered in the classroom and asked, "Did we make it?" "Are we going to fly on the *Vomit Comet*?" Every day that we didn't hear if we were selected, I was met with dramatic groans. I explained that NASA works slowly to ensure they select the best.

A couple of months later, I received a letter from NASA. From the hundreds of applications across the country, Sho-Ban Tech was in the final one hundred applications. Twelve teams were going to be selected. I was excited but nervous. I had made my point about succeeding to my students, my science colleague, and myself. We could, and did, compete in an open science and engineering competition. I decided to let NASA know that we were a small Indian high school, not a college. I wanted to show the students that honesty does matter, and we did not want to exclude a deserving college team that wanted to fly in the *Comet*. I wrote a letter to JSC on school stationery and informed the selection committee we were a small rural high school, and not a college, so please consider us out of the competition if only colleges are eligible to apply. I thanked them for their time. I sent the letter by certified mail.

I was very happy to have progressed this far with my team. We competed as a high school team against national universities, and we made the first cut. Our science was evaluated not on the color of our skin or pity, which sometimes happens with misconceptions of Indian people. Our team had a good research question. This is how excellent science should be done. I like to think of this as science equity. I called my friend and told him that we won a spot with NASA, not because we were Indian, but because we had good science ideas. We won without special consideration, treatment, or pity. This felt great. He agreed we won and he paid me the dollar he owed me.

A month later, I received another phone call from the JSC. My team had won a spot on the KC-135A! Our team was to build our experiment, get flight physicals, and report to Houston in a few months. Since I had not heard otherwise from the NASA officials, I assumed we had been approved as a high school to fly. I was told by a NASA official that the following schools were in our class of selected experimenters: University of Michigan, University of Texas, Austin, University of Utah School of Medicine, Lamar University, University of Washington, Utah State University, Washington State University, Rice University, Emory University, and Sho-Ban Tech!! I had a permanent big smile for days . . . still do as I think about it!

So, our team prepared to go to JSC. We built our experiment. We completed our flight physicals; we received permission to travel to Houston from our school board. We had fundraising activities for our team. The community, both reservation and non-reservation, supported us in a big way. They had special dinners for our team, much like the Mercury Seven astronauts. We had special TV, radio, and newspaper interviews. We were on our way to Houston—or so we thought.

CHAPTER 5
A Community Experiment

> *Throughout history, all human groups have depended on careful observations of the natural world. If they learned from these observations, they adapted successfully. If they did not, the consequences were probably deadly.*
>
> —Fikret Berkes (2012, 76)

Winning a spot to fly on the *Vomit Comet* means an invitation to join the NASA family, but it doesn't mean that NASA will build the project for you. Your team must build its own experiment and make it work. The team must build the experiment from their own design and engineering materials, and follow power, weight, and safety considerations regarding materials. All materials are required to meet the strict NASA guidelines to be durable enough to fly on a NASA aircraft. All tasks are not easy to do. Fortunately, I knew I had the brightest, most creative and fun students in the solar system. I believed in my students, and knew we could do great things together. By this time, we already had the experience of doing great things together.

In our science lab on the reservation, the students, staff, and mentors put so much effort and time into building our little experiment that we gave the experiment apparatus a name. The students called it Baby. Indian people give names to all sorts of objects, including people. Some are funny and some give a humorous description of the item. Humor is a big part of the Indian culture. Laughing at things, including people, is what Indians do.

SCIENCE CLUB

After our proposal was accepted, students were hanging around in my science room all hours of the day, including after school. I decided to see if we could do some science things. After all, the sports folks met after school, so why not a science club? I asked some of my

(*Above*) NASA Science Club. Photo: Ed Galindo.
(*Below*) Experiments done at NASA Science Club. Photo: Ed Galindo.

the answer for a while, and after about a week, I asked them again one afternoon, "What do you all think about an after-school science club?" The students had some questions of their own that I needed to answer first: "Will there be food?" "Does the science club cost anything to join?" "What kind of food will be there to eat?" "How do we get home after the club meeting?" "Will there be candy?" "Can we get extra credit?" "What kinds of things will we do?" "Will soda drinks be provided?" "Can we do this club and sports?" "Can my little sister or brother come?" (Since some of my students babysit.) "What kind of food will there be, and when will it get here?" All great questions. I found that food is a good thing to have for mentors and students in an after-school program.

I needed administration approval for the club. So I informed my principal that I was interested in forming a science club. He asked, "How much will it cost?" Before he granted approval for the club, I was told I needed to present this idea to our school board. I agreed.

School boards are full of fun folks. However, an all-Indian school board is really fun. This is because they are all hard working, and they all come from the community. It is a hard job to be a school board member. After working all day, they meet to discuss and vote on difficult issues. I presented my idea, and the board formulated the usual questions: "How will the students get home?" "Can they get extra credit?" "Will there be food?" I smiled and said, "Most certainly there will be a snack. Come on over and join us sometime." As it turned out, many did come for snacks and science.

We formed our science club. We initially named it "The Science Club," but the title sounded so nerdy. Who wants to hear this on the school intercom: "Science Club is meeting today"? If you are trying to be a cool teenager, then a cool name is wanted and needed. After much student discussion we decided to call ourselves the "NASA Club."

The NASA Club was open to all students at the school. The students who joined the NASA Club were motivated to be in the club. Many were good math students, some excelled at science, and others just wanted to be part of the club because it sounded fun and it was something to do after school. As time went on, the club size grew as the students started talking about the "cool NASA club" that did exciting, fun things. No one had to be a "superstar" to be involved. They just had to sign up and help where needed. Eventually the students who were not involved in science learned cool science things from one another.

For me, it was really fun to watch the students interact in an environment outside the classroom. I loved watching them share ideas. We also invited parents, students' little brothers and sisters, school board members, and alumni. All were welcome to join and eat some snacks.

The reservation community is small. After a while, everyone knows everyone else. It truly is a family of folks who see and know one another. This is how it was in my classroom. My students knew me and I knew them, as well as their mothers and fathers, grandparents, aunties, uncles, and cousins. I knew them because I visited with them, had dinner with them, laughed, cried, and socialized with them. They were my family, and, for some, I was theirs. As with all families, sometimes there are disagreements. This is how life is on the reservation. The point is, I knew the families well, which was a huge asset in classroom management. I could always say to a student who was misbehaving, "What would your grandmother think about this?"

Eventually, "NASA" for us meant "Native American Science Association." We liked this name. Later on, the use of this acronym was approved by the director of NASA himself, but that is another good story.

MENTORS

I believe we never accomplish anything by ourselves. We are brought into this world with help, and I think we leave this world with help as well. In between, we rely on help from others, and hopefully give help to others. This was the case with Baby. During the school year, I invited all sorts of potential mentors into my class from both on and off the reservation. They were good folks who were Indian and non-Indian, and they all exhibited two main characteristics: They cared about my students. And they possessed good hearts.

One of my class's mentors was an engineer. I will call him Dave. Dave gained experience with NASA as an engineer working on different

Mentors. Photos: Ed Galindo.

fly contraptions into space. He also knew Indian students' issues, as he is a Sho-Ban tribal member. However, I have found that when working with Indian students, once the trust bond has been made between the person and the students, it really makes little difference if one is Indian or not. What makes the positive difference is if the heart of the person is one of trust, caring, and compassion. I have found that what we see on the outside of a person may not be what the heart is all about (both good and bad). Truly believing in the students with compassion, understanding, and high expectations, *not* pity, is what Indian students need as part of a successful formula for education.

Dave was excited when I told him that we won a spot to test fly an experiment on the KC-135A. He realized what it took to be invited into the NASA family. As our class mentor and engineer, he gave us his opinion on how to build our experiment and how much it would cost. Dave asked the basic questions any good engineer would ask: How much weight do we have to work with? How much power? How much space do we have to work with? What do we want to accomplish? What is our timeline? Could I get him needed information today?

The NASA flight project was in addition to my full-time job of teaching. I was teaching six *different* science classes: Earth Science, Physical Science, Chemistry, Physics, Biology, and Life Science, all with labs. This kind of hectic schedule is pretty common with many small schools. Those who do not teach, like Dave, may not understand what this means. It is like having six jobs and six more with labs, for a total of twelve jobs to do every day, all at once. I have long ago concluded that teaching is not for the faint of heart!

FUNDRAISING

Over a period of a few days, the students and I provided Dave with the information he requested. A week into the project Dave called and said that he wanted to share some information with my team of student researchers. Dave knew about our NASA Club and wanted to meet with us after school. We were eating our snacks and having a great time, meeting and eating, when Dave showed us his colorful schematics and plans. Sarah, one of the students, asked the question, "Mr. Dave, what is this going to cost?" Mr. Dave said, "Well, Baby will be made from pure aluminum, bored and welded out of a single big block. For maximum strength and weight, I think we should be able to do this for about $20,000." I was eating a chip and started to choke. I wasn't sure if it was chip or the $20,000 figure for building Baby. Anyway, some of the students asked if I was OK. I replied weakly, "*No.*"

NASA waived the initial cost of test flying our experiment. This was part of the prize we won when the teams were selected. For a "gee-whiz" fact, my team was told that when payloads (also called experiments) fly in space, the cost is about $10,000 a pound. Since Baby weighed in at a whopping eight pounds . . . well, you can see what this would cost. Since this was a student experiment, that cost was waived. However, the cost of flight physicals, room and board, and transportation to JSC for a team of ten members was NOT waived. As my team and I did the math to fly us from the reservation to Houston, room and board for three weeks, and ground transportation, this worked out to be about . . . $10,000 a pound for my team to go to Texas. I was amazed. Most science departments on any campus, college or high school, are in a constant hunt for money. The Sho-Ban Tech Science Department was no exception.

When I started teaching on the reservation, I asked my principal a fair question: "How much money does the science department have in its budget?" The principal, a good man, said, "Let me look, I have the spreadsheet right here." He thumbed pages and double-checked with a puzzled look on his face. I thought, this is a good sign. He cannot believe what he is seeing so I must have such a large amount. Finally, after a long pause he said, "Looks like to me the science department is about fifty dollars in the hole." He said, "Well, that should do it and good luck." I was starting my science teaching career with a negative number. "Well," I said, "at least it is not one hundred and fifty dollars." We both shared a nervous laugh and I left thinking, now what?

I concluded that money, or lack of money, should not stop good teaching or doing things for students. I have seen many great teachers who personally fund the students they are working with. They are my heroes and sheroes. What I am talking about is the way many dedicated teachers go about helping their students with no fanfare or expectations for themselves. I have seen teachers find resources when none existed. I have joined my fellow teachers in raising money for things like coats for kids, food for families, and bus fare, not to mention endless classroom supplies like pencils, pens, paper, and notebooks. Teachers all across this country do this every day. An education budget that is short or nonexistent might not mean much to those away from students. However, it means everything when I am the teacher in the front of the class looking into the eyes of the students. My task was to find a way to fund our NASA experiment. I have personal knowledge that babies are expensive, both human and mechanical ones like the one that we just created in my science classroom.

As I stated before, we really do nothing by ourselves. So, I decided to try the reservation "tennis shoe experiment." This is how it works:

Let's imagine that a student on the reservation wants to go out for the basketball team. She tries out, runs hard and fast, shoots well, great on defense, does all the right moves. She is a star. She has made the basketball team. The school provides the uniform, but not the shoes. Basketball shoes are very expensive. For this example, let's use a figure of $150. Now, if you are on a tight budget, $150 is a lot of money to spend. So, what do you do? A family member will call on several other family members; the uncle or auntie will be called and a family visit will happen. "How are you doing? What is going on?" The usual family talk. Eventually, after a circular conversation, the conversation will come around to basketball. "How is Jennie doing on the team?" the uncle will ask. "Good," says the mom in response. "The cost of shoes is sure expensive," says the uncle. "Yes, they are, how about donating to the shoes." "Okay, how much do you need?" asks the uncle. "How much do you have?" says the mom. Finally, an end point is reached. "Okay, I have $20 dollars I can give to the shoes." This conversation is repeated and repeated to the community until the sponsor gets enough money for the shoes. Donors include cousins, uncles, aunties, and many other family members and extended family members. I like how this ends, something like, "Thanks for your help; by the way, come see what you donated for on Friday night at the school gym. Game starts at 7:00. We will save you a seat."

This is an example I have seen time and time again on the reservation. No one has everything, but everyone has something. The idea is to share what you have for family and community. This model is also good to use if the balance in one's science budget is a negative $50.

When Dave told us that for only $20,000, we could build, test, and fly our experiment, we had a classic example of what I like to call a "mentor mismatch." This happens a lot in schools. On one hand you have folks that want to help you and your students; on the other hand, you have the real situation with school finances. For example, I was invited one day to a local classroom to share what a science teacher does for a living and what I was doing with my students. I was going to share Baby's story, and our research program with NASA, and I was prepared with a slideshow, the works. Just before I started, a little boy's hand went up. I said, "Yes." He said, "Hi, I'm Billy. Did you know that this morning my goldfish, Goldie, died?" He was trying to be strong and not cry in class. Immediately twenty-nine other little hands went up. Then it started. "No, I did not know that," one girl said. Then another student, "How did this happen?" Another question, "Why was Goldie sick?" On and on it went for five to ten minutes. As a teacher and guest of this second-grade class, I knew what was important at the moment, and it was Goldie, not my talk. So, we talked about the great circle of life. Things are born, they live, we admire the beauty and great gift of life, then things die. My point is that mentors need to know what is important to the students. In the case of my second graders, it was Goldie the goldfish, as this was part of the student's concern. How Billy was doing was all of our concern. Billy and Goldie were more important at this time than my talk. I knew this and went with it.

I told Dave about the school finances and he understood what we were trying to do. The students ultimately found a lot of ways to fundraise. They worked the school concession booth during the football and basketball season. They sold pizza slices at lunchtime for $5.00 per slice. At Easter time, they sold decorated Easter baskets. They sold raffle items, like cool beaded items. I was impressed with how they could earn money. They raised enough funds to go to Kennedy Space Center to see their experiment fly. Fundraising was another good learning experience for students.

Mr. Dave became a good friend of the Sho-Ban science team. He observed and offered advice on all their experiments. He had worked for NASA as an engineer, so he had loads of suggestions. When the team flew the mission with NASA, Mr. Dave and his wife were invited to join the students in Florida at Kennedy Space Center.

CHAPTER 6
Birth of Baby

Ultimately science is story telling for understanding the natural world . . . "Coming to Know" for education in Indian tradition, is the best translation . . . There is no word for . . . science in most Indian languages.

—Gregory Cajete (2000, 80)

When a team is invited to fly an experiment on any NASA vehicle there are some basic questions that are generally asked: Is it safe? How much does it weigh? How much power does it need? What is the scientific merit and question of the experiment? Does it have to fly in space to have an answer to the science question? All are valid questions. Baby's official mission was to extract phosphate ions from the finely ground phosphate ore, mix it with water in microgravity, and attempt to make liquid fertilizer. We told the NASA folks that this experiment was designed to test the effects of microgravity on our method of extracting the phosphate ion from phosphate ore. Our experiment would also test general chemical and engineering processes in a microgravity environment. In our lab classroom on Earth, the extraction of the ion is easily performed and understood. It was my team's desire to perform the same process in the microgravity environment of the KC-135A that simulated space. Our research question was to determine if the mixing chemistry between phosphate ore and water was the same in space as on earth. Baby was to mix phosphate ions with water in a thirty-minute time frame. Baby had to do this by using its own on-board computer to push and pull actuators that made the water pass through the phosphate ore and a series of filters to make the phosphate-ion water that would improve growth of plants with our "space fertilizer." If this experiment worked, Baby would start with about one liter of water and produce about seven to eight hundred milliliters or about 75 percent of the original quantity.

Baby could be built out of an all-aluminum block for the low cost of $20,000, but not in my rez classroom. Baby did need to be built, though, and by the students, staff, and mentors at the school, not in some well-funded government facility. So I thought about the requirements to fly Baby. Our Baby had to be strong to survive the rigors of the NASA test flight. What we needed was inexpensive, strong materials, materials that the students and I could get at a reasonable cost. We needed strong, low-priced parts to hold the phosphate ore, water, and collection bags.

I asked my students to think about what materials would be strong yet inexpensive to get. This is where my students could shine. We asked, "Where could we find parts to build our experiment?" My students were experts at finding good quality but inexpensive things to "get by"! Reservation life is interesting in that residents use what is at hand for many things they need. Many reservations have a boneyard where old car bodies and car parts, hot water tanks, trailers, tires, and other peoples' junk are a gold mine for people looking to fabricate stuff from old parts ("bones"). One only has to think outside the box.

For example, say someone wants to create a woodstove. One just goes to the boneyard to find old discarded barrels, weld them together, cut an opening for the wood and the stove pipe, and weld on some legs. Voilá! Another example of recycling is passing down outgrown winter coats to younger members so they are used by more than one family member. The idea is to keep warm. Money is also scarce, so one finds things that are inexpensive and available for purchase with the money at hand. We had a rich discussion about where to locate parts that were cheap but also effective so that they could meet the rigors of NASA tests for flight. The stuff we needed would not be found in the boneyard. We talked about using truck brake lines for fluid transfer, model airplane actuators, and plastic tubes that were used in sprinkler parts.

"How about plastic parts?" Jake said. I said, "Okay." I told them that I thought we needed a cylinder shape to hold the phosphate ore, and tubes to carry the mixtures from the ore and the water solution to a collection bag. The students and I had several days of discussions about what sizes, shapes, and materials to use. I reminded the students that the parts had to not only be affordable, but also, new—off the shelf. The materials had to pass the rigors of NASA flight testing. No worries for a bunch of creative Indians.

After a few weeks, we had some good ideas. Our holding tubes would be sprinkler tubes from a local hardware shop. Our water lines, which would carry the mixture of phosphate and water, would be made from truck air brake line parts. The mechanical parts to push and pull the material through the lines would be actuators, devices that model airplane enthusiasts use. The collection bags we would put the fertilizer solution into would be made from hospital IV bags. The entire experiment would be bolted down in a case that would be made of Lexan, a shatter-proof material. To run our experiment, we needed to build an on-board computer. That required a computer program to be written by the students and mentors. The total weight would be less than eight pounds. So, our Baby was to be made of remote model airplane parts, sprinkler parts, truck airline brake parts, a couple of hospital IV bags, and some duct tape (but no bubble gum). All the parts were flight approved for safety at JSC when the NASA flight engineering and safety review teams passed Baby. We were now in the space research business.

Baby needed to be small as experiments go (about two by three feet, with a "birth weight" under eight pounds). I thought this was doable. At least we could see the size we had to work with. The real problem was the power. Our team had a maximum of about sixty-five watts of power to run our experiment. This is the same amount of power contained in a small light bulb. We had to develop an experiment that was self-contained, meaning it can run without humans doing anything but supplying power via a switch. The space shuttle astronaut's job would be to flip the switch, which could be an issue. How could our team solve this issue? Could we develop a simple process of flipping the switch to turn on Baby? Through our discussions about how we could solve this problem, I was demonstrating to the students that they had to understand the questions being asked, then use their critical-thinking skills to find a solution. The students had lots of responses: What if we did this? What would happen if we did that? Could the switch be placed on the panel to be accessible for astronauts? Finally, the students came up with a solution that solved the design problem by turning on Baby with a wireless switch.

We needed help with the computer program to run Baby. I told the students that I knew some friends who maybe would give us a hand. The students asked who the friends were and I told them they were going to be our team's mentors. The mentors were meant to help answer questions, not do the work to design Baby's operating system. That was the students' job. Some students were interested in helping to write the computer program that we needed for the operating system to make Baby functional. Some wanted to work with mechanical parts and help build the experiment itself. Others wanted to help find all the materials that we needed.

After many discussions, the students and I developed a plan. I divided the students into groups with a sign-in sheet for parts. Some researched truck brake lines, and some sought out plexiglass that was lightweight and shatter resistant. I wanted them to ask how much the materials were going to cost? Most of the mentors donated the parts for the flight!

The point here was to let the students do what they want to do, not to assign tasks. The students all had their own special gifts and ways of learning. In the end, all the teams would come together and combine what they had been working on for the good of the experiment. This way worked well for me . . . and Baby!

Once the students and I knew what we needed, we asked our mentors for help. An engineer friend, Tony "The Tiger," knew how to write the on-board computer program to run Baby. The computer parts needed to be purchased, but at a discount price, by Tony. Some students who were interested in computer programming worked with "The Tiger." Other students who were interested in the mechanical part of engineering worked with a mechanical engineer named John Boy to build the case Baby would sit in. Other students met with mentors who knew where to get truck air brake lines and sprinkler parts for a discount. Many of the merchants the students interacted with were so nice that they donated the small parts to the project. Many who donated parts, or gave our club a reduced price, received thank-you letters from the students—so the students also practiced writing skills.

Finally, the students and mentors assembled Baby. They hooked up the power and turned on the computer system. Would Baby come to life? When the students finally pushed the button, Baby did wake up and she worked—at least with no phosphate or water in

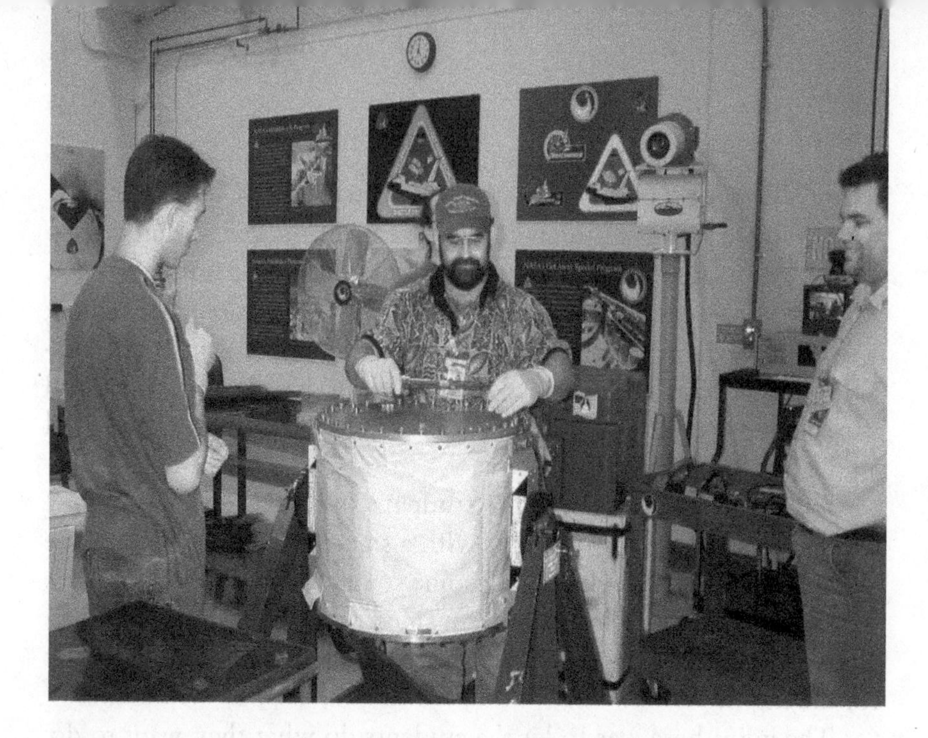

(*Above*) Jake, Baby, Al "The Hairy Indian," and Dave. Photo: Ed Galindo.
(*Below*) Jake, Baby, and NASA flight technician. Photo: Ed Galindo.

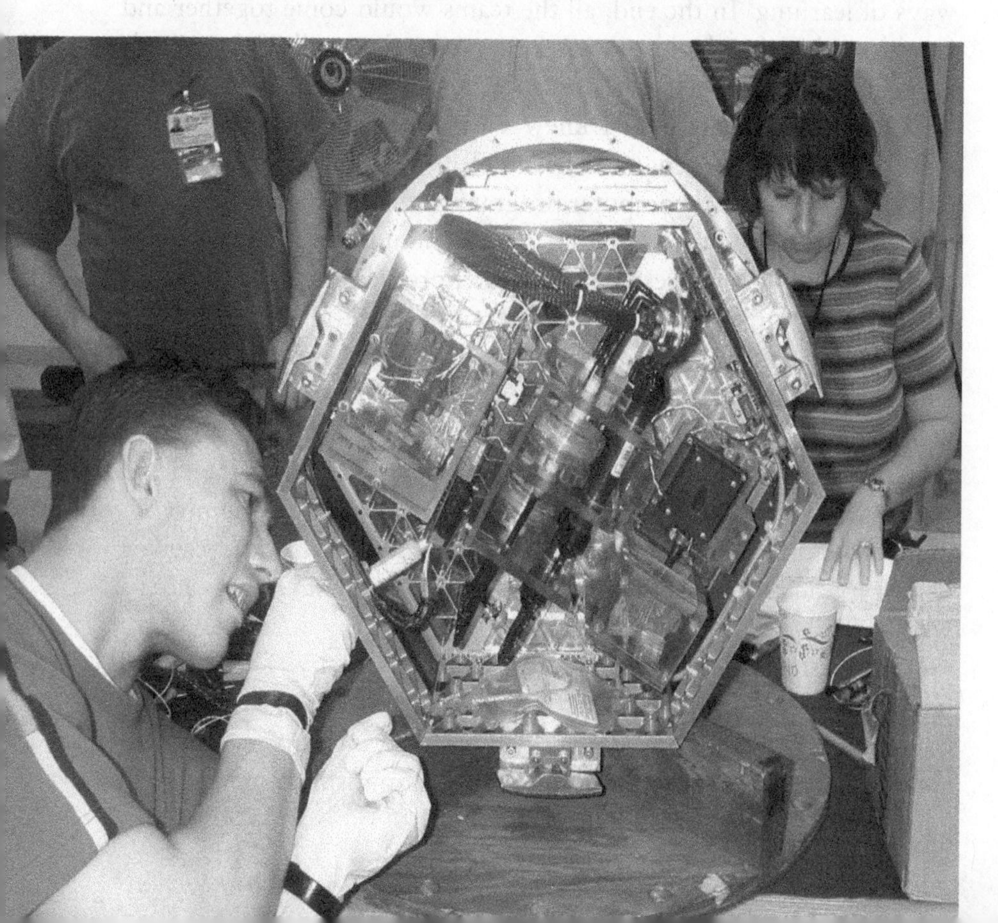

dry chambers. So far, we had met the power requirements and safety requirements of NASA, and the size and weight were right where they had to be for us to fly Baby.

Baby took on a life of her own. Baby was surrounded by a four-inch, C-shaped plexiglass case. Inside Baby's case we had placed two actuators, one on top of the other, controlled by a small motor that would push against a hospital IV bag of water. The pushing would force water down some tubes to mix in a chamber that contained our phosphate ore. This would answer two of our questions: Could we mix water with ore in space, and would it mix evenly?

Our team reflected on science, technology, engineering, and math (STEM), including chemistry and physics, as we learned to build an onboard computer to run Baby in space. We discussed the math of space shuttle flight, and the speed to obtain orbit. We debated the pros and cons of becoming an astronaut or an engineer working for NASA. We deliberated about growing food in space. We met people that sold us plexiglass and discovered some of these customers use plexiglass for making things bulletproof. The students held many newspaper and TV interviews. In the end, Baby was teaching our team a myriad of subjects in STEM, but also about teamwork and the long-term dedication that it takes to complete a difficult project on time and within a budget. Baby also taught my students the courage it takes to do a newspaper, radio, or TV interview and to think on their feet. I had come to appreciate our little Baby as a good teacher.

The total cost of the first space science experiment in our nation built by an American Indian high school team that would fly on an official NASA aircraft was about eighty dollars. None of Baby's parts were paid for from the negative fifty-dollar school science budget. Mentors and students raised the money we needed by working together. In fact, the science budget at this time was now officially a negative fifty-five dollars, including interest on the balance.

CHAPTER 7

Houston, We Have a Problem

An ecological person is a web of dreams of a rather superior kind that we can make into reality—if we act upon it.

—Gregory Cajete (2000, 63)

Finally, our team of Indian high school scientists was ready to go to Houston. We raised travel money through fundraising activities. We completed our flight physicals. We had a special good luck school assembly with an honor song for the team. An honor song comes from an Indian drum group that sings with the beat of a sacred handmade drum. Everyone stands for the song, and stands for the honor of the song, and why, and who is being honored. An honor song is truly a spiritual event. In this case, the song was sung to honor students and an Indian teacher who focused on STEM subjects. We were Indian space men and women. The honor song was for not just the students, but for our Indian family, school, and state. The honor was for the courage to attempt and accomplish a challenging idea. Honor songs are sung for many reasons, sometimes to honor sports stars, or for an athlete who has accomplished astounding deeds. This song was for STEM. Our society focuses on sports a lot, and it was good to see that same focus and honoring for academic achievements.

The team was ready to go. Then, the night before our plane ride to Houston, I received a call. My team was going to be disqualified. Yikes! The person at the other end of the phone line was from NASA and he was a bit excited. He was in charge of getting the teams onto the NASA base. He had noticed my team was a high school student team, and he told me that the regulations clearly state only colleges and college students can apply. He still did not know that we were an all-Indian dream team of science and engineering.

Now, there are a few things that make a science trip not fun, and one of them is being told the night before departure that your team is disqualified, especially after the Indian community has held a huge feast and had such a nice send-off. This is not a good thing to happen to your team. However, my students and I have had a few things like this happen before. For example, once my students and I were going to be arrested while we were eating our lunch on the banks of the Columbia River. But that is another story.

I asked a few questions. "Did we win the rigorous science competition fairly?" Answer from the phone voice from Houston Space Center after a pause . . . "Yes." "Was our experiment accepted on solid scientific merit and safe to fly?" Answer from the Houston voice after a longer pause . . . "Yes." "Did we win because we're a small Indian reservation school?" Answer from the Houston voice, "No, the team won a spot to fly because your team has a good scientific question and a safe experiment." Finally, the last question, "Did you read the certified letter I sent to the director of the competition, informing him that we were a small rural high school in Idaho?" I did not say "Indian high school." And answer from the Houston voice after a pause . . . "I guess not, when did you send it"? "Two months ago," was my reply. No reply from Houston. "Hello," I said, "Is this phone working?"

I believe that life truly is a compromise. I also believe that we are on the planet, Mother Earth, to help all things whenever and wherever we can. I believe that most people want to help do what is right. Sometimes they need a little push in the right direction. This was our worrying situation the night before our team was to leave for Houston. We were going to be disqualified because we were a high school, a high school that came up with a good scientific research idea, one that had been accepted by the high, rigorous standards of NASA and had won a chance to fly on the KC-135A, the elite training bird of NASA, and to meet other folks and hear about ideas from around the country. We won the right to fly fairly and squarely. This was a major opportunity for my team of students. I was going to fight for our chance to fly. My students were worth it to me. Two hours passed and I received a phone conference call from NASA with folks who were high in the NASA food chain. They asked if we wanted to come to Houston and fly. I answered, "Yes!" Did we want to make a compromise? I paused. "Maybe."

This was the deal. All team members from any university in the United States needed to be at least eighteen years old. *No exceptions*, as stated by the highest of NASA officials. Did I have students on my team that would meet the age requirements? Well, during the final months of the competition, I had thought about something like this happening, so I had made sure that—except for two of the brightest women students—all ten members of my team were eighteen years old. The students were experts about following the rules. I am not so sure their teacher was! The students were ecstatic they could fly, but miserable because the whole team could not all could fly. As a cohesive NASA team, we discussed why everyone could not fly. I reminded the students that we were a high school team and NASA had age requirements about who could fly and who could not. NASA had already shown us "good character" to let us get this far, especially since we were competing with all college teams. I felt this was the best we could do at this time.

Traveling from the reservation in Idaho to Houston, Texas, was a very big deal. We needed the entire team to get Baby to work well in flight. We hatched a plan. Those who could fly on the *Comet* would, and the two who were unable to fly would be part of the support team on the ground. They would help support the team. Everyone was needed. The NASA team agreed everyone was needed at ALL levels.

I told my students that I did not know how the experiment would do, but that we needed to try. I asked them if they were willing to try. They all said yes, so we would try no matter who was on the team.

I informed the NASA officials from Houston that my compromise was to bring all ten students. Two could not fly, but they could be part of the ground crew to get our experiment ready and support the students who did fly. After a pause . . . the Houston voices from NASA agreed. We reached a compromise. Our team was instructed to report to JSC and Ellington Field. The first, brightest, and the best American Indian high school science team was going to Houston to fly on the KC-135A with the best and brightest university student teams in the nation. We were excited!

I briefed my team of staff members and students on the following day about what had happened with NASA. The two women students were disappointed, but they understood the situation, and we talked about life requiring compromise. They would go and support the team. Just getting to Houston was a big adventure for all my students.

Imagine being from a rural setting, never mind being from a reservation, never having flown in an airplane. Some students had not traveled far from home, and if they did, they always traveled with family members. Many Indian folks have been raised in a large family atmosphere, and not just immediate biological family, but huge numbers of extended family members. On a reservation, one is never far from family and friends. The feeling of family is a constant presence. It is a good feeling. I was asking my students to go a long way from home with me, and trust me that we would return home safely.

My team was going on a journey, a journey far from home and far from family. Houston, Texas, may as well be situated on the moon or Mars. It seemed just that far away and unknown.

When faced with a difficult journey or task, some Indian folks turn to a higher power, the power of prayer. In our classroom we carried out a special Sho-Ban blessing ceremony in which a respected elder was asked to bless all of us. This type of blessing is not for material goods or winning a competition. It is a blessing of respect and compassion for all things on Mother Earth. I mean all things, the rocks, the plants, the water, all life in the water, and all life on the land—a prayer for compassion, for the understanding of all things and all people. The prayer is for all people and relatives around this world. Most American Indian prayers have a special ending. A prayer for all my relations. Now we were ready!

CHAPTER 8
The Mission

The top of a mountain is a metaphor for a place of perspective about where one has been, where one is, and where one is going.

—Gregory Cajete (2000, 41)

In April 1997, ten of my students and two adults—the school superintendent, Dr. Phillip Shortman, and our JROTC instructor, Army Reserve SFC John Moeller (retired)—traveled with me to Ellington Field, about ten miles north of JSC in Houston, Texas, to report to a hangar where the beautiful bird, NASA's KC-135A, was waiting to test materials, experiments, and people.

Once the team checks in, the pre-flight training and examinations begin. As each team reports to the hangar, initial paperwork is confirmed. For example, all team members must have a second-class flight physical that involves a thorough examination, including blood work. Since there was no NASA-recognized flight doctor on the rez, we traveled to Pocatello, Idaho, for this. The physical report is certified on site. Next, the team members, not the instructor, lead an operations and procedure briefing of how the experiment is designed to work. This briefing is presented in front of the NASA safety engineers and scientific committee. The job of these fine NASA folks is to ensure the experiment is safe to fly. NASA has extremely strict protocols for safety with experiments, especially student experiments. The job of the students is to prove that their experiment is safe to fly on the bird. If it does not meet NASA protocols for safety, then the team is not safe. If the team does not pass medical paperwork and operation mechanics, the experiment is not safe to fly. If the team cannot manage corrections, the team goes home. I have come to appreciate this about NASA. They have very high expectations, and they are enforced. If the NASA science and engineering professionals deem the experiment is safe, then the team progresses to the next phase, physiology training.

Physiology training involves classroom training with instructions on hypoxia, decompression sickness, and one of my favorites, spatial disorientation, also called motion sickness. The classroom work culminates in a simulated ride in NASA's altitude chamber. NASA personnel trains the team on emergency procedures and what happens when the body experiences hypoxia. For example, at a simulated effect of 25,000 feet, team members take off the oxygen masks for about five minutes to simulate what happens to the body without sufficient oxygen to the brain. After about two minutes, some team members exhibit strange behavior. Answers to simple math and thinking questions do not make sense, but when questioned afterward, participants in the chamber will swear that they felt as normal as anybody.

After successfully completing physiology training, the team must pass a written test on flight physiology and basic flight principles. If the experiment is safe, and the team passes its entire test sequence, then the team is invited by NASA to fly on the KC-135A. It is a huge honor to be invited to fly.

The Sho-Ban Tech team was very excited, and the students thought I should buy them dinner on base this night. I could not have agreed more. Phillip and John and I decided to host an honor dinner. We learned that many Indians are working for NASA, in countless different jobs. Indian engineers worked on rockets, mechanics worked on the training airplanes, and others worked in the front offices. They were all excited we were there and helped to put this honor dinner together. Their family members sent food for the students and put the menu together. It was amazing to witness how all of it came together. During the dinner we feasted on salmon, bison, fry bread, corn, squash, of course the ubiquitous Jell-O salad, and other traditional foods. Prayers were said. The drummers sang an honor song. Indians and their drums have been intertwined since time immemorial. The drummers for this event were also already on site working for NASA. They knew why this dinner was being held, and knew the appropriate songs to sing. Some of our NASA students were drummers and singers as well, so they joined in and helped sing. A number of drum groups sang honor songs and others sang victory songs. It was a celebration! Northwestern tribes call this kind of celebration a potlatch, where everyone contributes something.

The entire process of safety review, classroom physiology, and exam passing takes about two weeks. On week three, we were ready to fly. I was proud of my team. They passed every test that NASA asked of them. Sho-Ban Tech, the all-Indian team from Fort Hall, Idaho, was ready to fly in the *Vomit Comet* with the best of the best. We had the right stuff!

FLIGHT NUMBER ONE

In the morning at 0900 hours, we had a final briefing with NASA and the crew of the KC-135A. We would board with six other teams. Baby would fly. Always in the mood to create their own fun, the students brought a small soft football to throw around in the rooms where we were staying. Jim, who was raised by his grandmother in a traditional manner, asked, "Can we bring this football to throw around during flight?" I said sure—it would help them refocus their minds, have fun, enjoy the flight, and NOT be focused on vomiting. This was an important concept of the flight; one of the astronauts told me about this trick of refocusing the mind on some activity and it worked wonderfully for the Sho-Ban students. Many of the other college teams took anti-vomit drugs, which was their choice. The Sho-Ban students took their small football to pass to each other during the flight. Our students knew firsthand what drugs can do to the body. Not one of our team members got sick! I advised them, "Tomorrow enjoy yourselves; imagine this is a very large roller coaster ride. You all like roller coaster rides, don't you?" They all nodded their heads yes and I smiled from ear to ear. I said, "Now that's the kind of fun I'm talking about."

The next morning our team was up early and ready to go. We had a very light breakfast and reported to the hanger where the KC-135A bird was housed. The flight team of the KC-135A led the morning briefing, including what the weather was like, what our duration of flight time was, and the most important part, where the vomit bags were located. The usual stuff one gets as a team about to ride on the *Vomit Comet*. During the discussion session, I was a proud teacher watching my Indian high school students ask questions and engage with university students from around the country. One team dyed their hair blue for this ride. My team called them the "blue smurfs." A spirit of fun was developing. Baby was brought on board. She was locked down, and her power cables were hooked up. The flight team would turn Baby on during microgravity conditions, or when they were floating. This would give the best-case scenario of what we wanted to know. Would Baby work in space to make fertilizer in microgravity?

The team was in place, Baby was ready, the bird was ready. The NASA medical personnel were asking if the teams wanted to take drugs for motion sickness. All teams took the drugs except our Indian team. A staff member and I were flight ready, but we had elected to give our seat to students as only eight could fly. We would fly four students one day, and four students the next day. The adults were backup in case someone was sick or did not want to fly. I wished the teams good luck and left the plane to wait in the hangar with the rest of the ground crew. We would be able to hear, but not see, what was happening during flight.

The bird lumbered down the runway and quickly climbed into the sky. Gaining altitude, it made a slow left turn toward the Gulf of Mexico. I said a silent prayer for the safety of all the teams. For ninety minutes, the aircraft did about thirty parabolas in the air, which are similar to roller coaster movements. I could hear my students doing well, having fun, laughing, joking, and really enjoying themselves. I wondered about them, and Baby. I was anxious for them to land. After ninety minutes over the Gulf of Mexico, they proceeded toward home base.

I watched as the KC-135A touched down. It is a massive white plane landing with wing flaps down on the runway. I noticed a large yellow water truck drive out onto runway toward the KC-135A. The truck has water spray hoses on it that are used to wash down the *Comet.* Specifically, to wash the vomit from the inside of the aircraft. Science at times can be a stinky, messy business!

Not one of the Sho-Ban students vomited during the flight. Everyone on our team had had fun and was learning to float in microgravity. However, other team's members, including many of the blue-hair folks, had multiple issues with vomit. Jim gave me a picture later that day that showed a blue-haired student with a nice curved arc of vomit about 1.5 inch in width, as it floated along in microgravity. I imagined the smell.

The students had some troubling news. Baby did not appear to function. We had another flight in only twenty-four hours and had to repair the problem pronto. Once I knew the students were okay, I focused on Baby. What had happened to Baby? The team had run the test sequence many times in our lab at school. Our team took Baby back to the hanger to inspect her. Quickly, we took Baby apart and right away we could see the reason. Baby was plugged up, or "constipated." This happens to babies once in a while. The cause of Baby's constipation was a filter problem in our secondary chamber area. Sarah asked, "Could we replace the small diameter paper filter with a larger pore size?" Where could we get a filter that had a larger pore size? I looked around and saw the flight crew briefing area. They had served us coffee earlier during the day in this room. I decided to try an unused coffee filter from the briefing office. When I asked a secretary if I could use a filter, she looked at me a bit strangely and said, "Sure, use whatever your team needs." We installed the coffee filter and tried Baby again in our hotel room. She worked great. Baby was no longer constipated. Baby was making fertilizer. I was happy, because I know firsthand that having a constipated Baby is no fun for anyone involved.

FLIGHT NUMBER TWO

Our second team of Indian microgravity experimenters was ready to try a ride on the comet. They had heard all the stories of how much fun it was, and we were all curious to see if indeed Baby would run through its program and *not* be constipated. It does not get much better for an Indian science teacher from an Indian reservation than having students excited about flying, excited about the science, and watching a Baby make fertilizer.

As on the previous day, we went to a briefing. NASA offered nausea meds for the team if they wanted. Our team declined the offer, and then boarded the aircraft, strapped Baby down, and plugged in the power cables. The rest of us said our goodbyes and went into the hanger area to wait.

On this second flight, the ground crew could again listen to the interaction of the students. I was anxious on two fronts: one, I wanted the safe return of my students; and two, I wanted Baby to work properly. Again, after ninety minutes in the air, the KC-135A headed back to home base.

The bird made a beautiful landing. After she taxied close to the hangar, we went aboard and gave our students hugs, and were told by the Indian fliers that Baby worked perfectly! The students were so excited that they were able to repair Baby so the experiment worked, and again no one on the team vomited during the flight. Baby had completed her mission. She made fertilizer in a microgravity environment. Not only had the first all Indian KC-135A flight teams *NOT vomited*, but our Baby had made liquid fertilizer. Man, that is good stuff to have happened in Houston. I was smiling so much that my jaw hurt.

HOME

It was time to leave our friends at JSC and head back home to our school. We had been away for about three and half weeks. At the end of each day when the students were not in NASA training, I told them and the other adult team members we could visit the sights around Houston. I suggested students go in pairs if they wanted to explore. As a small tribe, we explored the area museums, library, the university campus, and the butterfly garden. Naturalists at the butterfly museum lectured about the life cycle and how delicate wings were made of fine powder. We also learned that the butterfly flies in silence.

Houston is a city of 2.31 million people with an area of 669 square miles, so the students had to use a map to plan their trips. However, even with the maps, we all got lost in the big city. We knew where to go, so when we got lost, like getting lost in the forest, we just backtracked, looked for landmarks, and went back the way we came. Sometimes we asked clerks at gas stations where to go. The people we asked spoke with heavy Texan accents and often were difficult to understand. So we were even more confused about the route to take.

Just seeing a huge city when you live on a small reservation is a learning experience. The students asked lots of questions: Where do people find all the jobs to work here? How much does it cost to live here? Why would people construct a butterfly museum? All good questions. We talked about them all. For example, residents of a big city such as Houston may never have gone into the woods to see insects, much less butterflies!

Of course, we had a visit to the edge of the ocean. Since Idaho is a landlocked state, the rhythm of the ocean waves was mesmerizing and magical for everyone. As we stood on the edge of the pounding surf, we looked out at the horizon and discussed how animals could have come to the land from the ocean in prehistoric times. How did they adapt themselves to breathe oxygen from the air? We pondered the lifeblood in our own veins containing elements from the ocean. "Did you ever taste your own blood when you were cut? It is salty like the ocean."

The students got a bit homesick, but they found we were our own tribe that looked after each other. Returning home to Idaho and the rez, we were greeted with a great reunion of smiles and pats on the back. In the following days, the students told their own stories of science, flight, NASA, the *Vomit Comet*, and how much they learned not only about science, but life in Texas and how much fun STEM could be. I was proud of our teams, both students and staff. They did a wonderful job of representing their Sho-Ban Nation, family, and themselves.

When we returned to Idaho, we reflected on what we accomplished. We took time to re-honor our students, mentors, parents, uncles, aunties, grandparents, and team. The whole community made some sort of contribution to the project, whether it be time, food, or words of good support.

We also held a debriefing in our classroom. We shared what we learned with the other students, and we talked about what we wanted to do next. I wanted to know how clean the samples were that Baby made. Was it good enough to water plants and support plant growth? Some of the chemistry students and I took the sample that Baby made on the KC-135A for a chemical analysis with one of our research mentors, who worked with phosphate and water. He is experienced in examining these classes of samples, but he'd never examined one from a *Vomit Comet*.

Rob, the research chemist, wanted to meet with our NASA team. So he came to our classroom. Rob was in his mid-forties, had a beard, and was dressed in jeans and a cowboy shirt. He looked like a regular guy, not like a white-lab-coat, wild-eyed-scientist. He gave our NASA students the chemical analysis results. The samples were clean and good enough to water plants. "What does that mean?" Sarah asked. Rob explained that the sample Baby made on the *Vomit Comet* could be used for plant watering without harming the plant with an excess amount of phosphate compounds. This meant that our experiment worked well.

I was pleased with Baby and the space fertilizer she had made, but prouder of my students and the mentors. We were doing more than answering questions in the back of the science textbook. We were applying science concepts to life and testing those concepts with the best of the best with NASA. We were working on some interesting research problems and working toward viable solutions. My students were gaining more and more confidence in what they could do, and they were less afraid to talk to other people about what they are doing. They started to see a world far beyond our classroom walls on the reservation. Teachers, mentors, and community members were now coming to our NASA Club after-school meetings. The STEM work we were doing involved chemistry, physics, electronics, and computer programming, as well as learning to work with other teams from around the country. For many of the students, science and STEM were becoming as fun and important in their lives as the sports program. The learning became something our team did to make Baby fly. They had a reason to learn STEM. They all did a great job!

CHAPTER 9

STS-91 *Discovery*

In Indian creation stories, humans and celestial beings interact . . . sky beings come to Earth and humans and animals visit the sky realm and participate in the course of each other's lives.

—Gregory Cajete (2000, 217)

I was back at home in my classroom doing what I enjoy doing, learning and teaching with my students, when a visitor arrived in my classroom. It was Al, "the hairy Indian," the government research engineer. I thought to myself, "Uh, oh."

Al was excited to see me, so much so he drove about an hour to visit with me rather than calling. By this time, we had become good friends, so I liked seeing him. He had some news to share. Al followed our progress with *Vomit Comet*. He was excited to tell me that the Rocky Mountain Space Consortium (RMSC) was going to participate in a space shuttle science experiment with students called a "Get Away Special" (GAS). "What did that mean and why is this important to me and my students?" I asked Al.

Let me tell you why NASA is an interesting group to be associated with. NASA has very high expectations for STEM experiments that fly in space. NASA is a world player with folks from around the world who are involved in the process of space flight and space-related research. I wanted to test my research ideas, but I had another reason for being involved with NASA: it needs more diversity. For years it was, and still is to a large extent, an elitist club. When the space flight program started, only short, white test pilots with crew cuts could be part of NASA. And only "bold, type A personalities" could fly with NASA. Today, that image has undergone a remarkable transformation. Women and individuals with all ranges of skin colors and ethnicities are not only flying, but building rockets and teaching, and are involved in every facet of NASA life. This is good to see. It was a long, slow process, and one that still needs to be pushed along.

I found that NASA does not lower the safety bar on experiments for any reason. If the NASA experiment is safe, and asks valid space research questions, and your team can pass all the tests for flight, your experiment flies; if not, your stuff does not fly, period. I liked that idea then, and I still do today. I wanted to also see if my students could be successful and be tested at this high level. I knew they could do it.

NASA knows things and intends to know more things. For example, they are interested in who was successful on the *Vomit Comet*, and who was not. Word circulated that our experiment was successful, and that, even without taking motion sickness medication, none of the all–American Indian student team was sick on the *Vomit Comet*. NASA wanted to know why. I told NASA we were just a "dream spud team from Idaho." I believe there's another reason. My students were having so much fun, they didn't get sick.

What Al was proposing is that we, Sho-Ban School as ourselves, not as Tech, enter another open competition to fly an experiment on a space shuttle mission. I asked some questions: Why us? Who was in the competition? How much did it cost? And was NASA providing any snacks? Al laughed and said, "We at RMSC think your team has a good chance in that you have Baby. Baby worked, and Baby is small enough to fit in the payload bay of a space shuttle. The power requirement is low. The materials are safe. Your team can produce safe scientific results, and most importantly your team can meet a deadline, and within budget." I told Al that he would have to ask my students what they thought about this new mission. They asked all the right questions: How soon would the experiment have to be ready? What was the competition like to fly an experiment on the space shuttle? Did our team really have the right stuff to fly again? And would there be any good snacks through this entire process?

FUN WITH PHOSPHATE ORE

My students and I thought about the idea of flying an experiment on a space shuttle. We had already completed the *Vomit Comet* mission, so we had some idea of what we needed to do. Some of the students who went to Texas had graduated, but they mentored the current students as alumni. After considerable discussion and thought, we decided to apply. After all we'd been through, as Jake said, "Hey, how hard could it be?" We soon would find out.

We found out that to ride on NASA's magnificent KC-135A *Vomit Comet* bird was one thing, but to take it to the next level of space flight was an even more difficult task. To be considered for a space research shuttle flight, the first step is to complete the application. NASA folks knew who we were, so I decided not to trick them this time. I would advise you not to do it either. NASA folks as a whole are not too fun to kid around with. As an Indian man who loves to joke, I am still working with NASA on that, and I have made some progress.

The application process was even more intense than for the KC-135A, and we endured even more difficult questions. For example, "Do any parts have outgas qualities?" Our experiment was a little Baby; of course, we outgas and urinate. "What is the origin, weight, composition, tension strength, and number of screws that will hold the experiment in place?" This is just a sample of the many detailed pages in the application. Once this list was completed, we sent it off to NASA at the Johnson Space Center. The list of questions and our responses would be the baseline for the visual inspection that flight engineers and scientists would perform to determine if we were safe or not. Flying experiments in space is demanding work and requires serious thinking but this is how it should be. Space is a very harsh, dangerous place to live and work.

We completed the application. I say "we" because my students and mentors had to help me complete the extensive application. I was a PhD student during this time and was working with undergraduate and graduate students at USU to fly experiments in space as well as teaching science to my own Sho-Ban high school students. It was satisfying to see all groups working together. We had some of our own alumni with us as well as the USU groups. All of us were working together. The important part here is that we *all* got to know one another as real people. Being academic partners was an opportunity to talk about higher education and the commitment it took to attend a university. The USU students became academic role models for the students.

However, my Sho-Ban students became cultural role models for the USU students and taught them about Indian culture, especially Sho-Ban culture. Most students at USU had never seen or worked around Indians. An estimated 70 percent of the university's student body reports a religious affiliation with the LDS (Mormon) faith. Their ideas about Indians came from the Book of Mormon:

> According to the Book of Mormon, Indians were descendants from Laman and were therefore called Lamanites. Laman was the rebellious son of Lehi who had left the Old World and sailed to the Americas in the seventh century BCE. In the Americas, the Lamanite had grown distant from the teachings of God, become fierce and warlike, and had acquired a darker skin color.

Forrest Cuch, in *A History of Utah's Indians*, writes: "From a purely ideological point of view, the Mormons believed that the Indians were a remnant of a people who fell out of grace with God, were given a dark skin as a sign of their spiritual standing, and who now lived in an unfortunate condition awaiting restoration to an enlightened state." Today, Mormons continue their missionary efforts among Indians. The missionary experience is probably more important to the young Mormons than it is to the Indians. Although my students were high school age, they had heavy discussions with the USU students concerning their beliefs about Indian students. It was a good exchange for both groups, which I appreciated, since I was attempting to create a balanced relationship.

The Sho-Ban students were going to partner with the Physics Department at USU on a program called GAS (Get Away Special) CAN because the experiments fit in a canister that is placed in the back of the shuttle's payload compartment. Mr. Gil Moore purchased one of the first "cans" for student use. We owed Gil a huge thank you. This seems like something a Baby would do . . . make gas in a can. GAS is a well-established program with USU, and we were honored to be a partner. We submitted our part of the paperwork. Essentially, we would fly Baby in space this time, and see if she could make fertilizer again.

A couple of months elapsed and Al came to see me again. As before, he was excited and I again thought, "Uh, oh." Al asked me if I had heard any news about our space flight. I said no. But he had news to share. My Sho-Ban team had won a spot to fly on a space shuttle!

We would be getting an official letter soon. Now I was excited. I told the students what had happened. Some asked, "Where is Kennedy Space Center?" or "How do we get there? Where do we stay?" And lastly, "Can we see the ocean?" The trip to Florida was as great an adventure as the experiment itself: traveling across the country in a plane, finding a place to stay, and looking at the ocean in addition to seeing our Baby go into space. This was a fun and exciting time for me and my students and the community.

We told our NASA students and excitement spread like a fire through the community and all over the school. Finally, I asked Al what the next step was. Al told my NASA students we had to pass a shake test. "What did this mean?" asked Jim. Al said that we had to get Baby ready to pass a launch test on a special test apparatus located at Space Dynamic Laboratory (SDL), a premier space testing and research facility close to USU. SDL would schedule a time when we would bring Baby, filled with a load of phosphate and water. The shake test would simulate the launch vibration and determine if Baby could withstand the rigors of a shuttle launch times two (NASA's standard is for projects to withstand anything times two). Al asked the students if they had any questions. After a long while, one of the students' hands went up. "Yes," Al said, as he pointed at her. Betty said very seriously, "Don't you know, Al, you should never shake a Baby." Al was stunned. The students and I laughed so much that we had tears in our eyes.

FLIGHT TESTS

Our team went to SDL on our scheduled test day. Our team's first test was to pass the NASA shake test. Baby was bolted to the shake test table by the students. The students then did a final inspection and nodded that they were ready for the test. The students stepped back, put on safety goggles, and joined the NASA professional engineers to determine if Baby would survive.

This was a new environment for my students. About to conduct a simulated space shuttle launch experiment, they were in an extremely high-tech lab surrounded by scientists and engineers. The experiment was being led by the students themselves. I was so proud of them and

Proud students, ready for the shake test. Photo: Ed Galindo.

Baby. We were ready, and given a countdown. 5-4-3-2-1. The SDL staff turned on the power and noted the time, as this is a timed test. Baby was "all shook up." The test simulated not only launch forces, but doubled for a longer duration. Baby passed beautifully. Nothing came loose and Baby still worked well after the test.

With the shake test behind us, and the science/engineering flight sequence approved by NASA engineers and others, we were now required to go to Kennedy Space Center (KSC) in Florida to integrate Baby. Baby would be given some more testing by the good folks at KSC. If Baby passed her tests, then she could fly in space. If not, she would return home.

In order to conserve our money, instead of trying to take our whole team twice in the same year, I decided that I would lead the integration team at Kennedy accompanied by some of the USU students. Because USU was our partner in the project, the USU students had to travel to

KSC as a prerequisite for their programs. Once again, I discussed with the students their roles: creating the experiment, finding the parts, and making Baby run. Each student group had a job. I was going to make sure it all worked. The Utah State students' job was to put Baby in the "CAN" at KSC. They also assisted with the integration tasks of welding a safety cage and helping with the final wiring to ensure the experiment passed all inspections to go to space.

When Baby passed her test, we would gather the financial resources to take my Sho-Ban students to view the launch at KSC. With very limited funds, this was the best we could do. The students understood my rationale. It was expensive to go to KSC and they had already seen all that I was going to do with the *Vomit Comet* ride in Houston. They *all* would go to KSC when we were ready to "fly" on Space Shuttle *Discovery*. The students understood and all agreed that this was a good plan. They had one suggestion for me: keep Baby safe.

The USU/Sho-Ban team had a week to pass all the final inspections at Kennedy Space Center. The integration room is the place where all space experiments are inspected with a critical eye to ensure that they are safe to fly on the spacecraft. In the integration room, NASA checks power requirements and that all parts are flight ready.

I became aware of how important it is not to fail at this level. If our team project fails the rigorous inspection and the inspectors do not sign off, another team is requested to fill our place. The integration room at KSC entertained teams from Canada, Australia, and American universities including Penn State and Utah State University, all ready to take our place. The flight inspectors are a very intense bunch. Their signature means that the experiment is safe to fly. Lives depend on their signatures. They check everything, and I mean everything, from the exact manufacture number of the screws to the exact make of the bolt. It is a very demanding, intense week of questions and answers about the experiment. The NASA safety team examines parts of the experiment that may fail and make the mission unsafe. One of the inspectors told me, "Experiments meant to be on board the shuttle are considered unsafe until proven otherwise." This is their responsibility, and they are excellent at it. The safety of this segment of the mission and the lives of the crew rest on their shoulders.

At the end of a stressful week of pressure, electrical, and hazardous material tests, the team passed the entire KSC inspection. It is a great moment when a NASA flight team of engineers and scientists sign off officially. The first Indian-built science experiment in our nation was given final approval to fly on a space shuttle. Soon Indians were going to enter the space race.

THE ALUMNI

The world of NASA space shuttle flight includes thousands of folks from around the world working together. It is really quite magnificent to think about and see. After we became part of the NASA family, it took us a year and a half to fly with a space shuttle. By that point, most of the students who had flown on the KC-135A had graduated. This is a very long time to maintain student interest, but I had a plan. I formed a NASA alumni club.

Space shuttle STS-91 *Discovery* at Kennedy Space Center ahead of launch, 1998. Photo courtesy NASA.

The NASA alumni were Sho-Ban High School graduates who returned to help get Baby ready to fly. They were still excited about what the club was doing. I invited them to come back when and if they were in the area. Some were working full time for the tribe and could not get away. Several would come with us to KSC, since they were helping us to raise money for the trip. They would be the mentors for the next generation of young Indian scientists. This concept of helping the younger ones on the reservation is not new for Indian people. Since time immemorial, Indian people have employed the style of learning in which older graduates/people help the younger students. This type of education involves everyone sharing the skills they have with younger students; it is a way of life.

The role of student alumni on long-term projects cannot be overlooked. They maintained enthusiasm for the NASA projects and kept the high school students motivated about NASA and space. Since some were in their second year of college, the alumni also served as role models for my high school students. Many of these alumni were attending Idaho State University, Spokane Indian College, or Indian colleges in Montana. Their majors ranged across environmental sciences, fisheries, forestry, computer software, and other STEM fields. Their aim was to return to the reservation to help their people. Some were employed by the Sho-Ban Nation and doing quite well. One of my students, Sammy Matsaw, went on to graduate school to earn a PhD. All my students knew who the alumni were because just a couple of years earlier, they were in the same classroom (with the SAME old teacher—ha). They were a big help with encouraging students to stay in school, and with fundraising, as I was busy teaching classes during the day. NASA and school alumni are very good sources of ideas for a science teacher!

A teacher's time is stretched to the limit every day. There is never enough time to accomplish everything during the school day, even when you include time after school and on weekends. Alumni can save your life.

BABY FLIES

The alumni club helped us to stay flight ready and maintain students' interest while we waited for a launch date. Finally, it came by email.

NASA's STS-91 *Discovery* would carry our Baby into space. Eleven students, along with Dr. Shortman and Sgt. Moeller, would travel to Kennedy Space Center as VIPs and watch our little Baby lift off with STS-91 *Discovery*. The launch was originally scheduled for March 21, 1998, but did not actually fly until June 2, 1998. It is typical for NASA experiments to be delayed for months, if not even longer. Weather, equipment failures, and safety requirements all play into whether an experiment flies. Students have to plan and raise money knowing that flights may be delayed. The NASA students knew that other NASA experiments were prone to delays. We all believed that someday we would fly Baby, and we had to work with NASA to fly. I had assured the students who had graduated that they could still be part of the team as alumni. Any delays would just give us more time to raise funds! The students were OK with the process if they knew why we delayed. As a teacher I was as transparent as I could be about this idea. We all waited—and talked nicely to the gate agents about refundable plane tickets.

On the day of the launch, Indian students, mentors, and Sho-Ban community members were at Kennedy Space Center in the VIP section. We all focused on a single goal that day—the safe launch of STS-91 *Discovery* and watching our Baby fly.

It was time to launch. 5-4-3-2-1. And a cloud of white smoke went up. We heard a thundering roar. Even though we were five miles away, we could see the shuttle on the pad. We could feel the vibrations from liftoff in our chests, which were followed by a loud boom. We watched, mesmerized, as the announcer said, "The shuttle STS-91 *Discovery* lifts off." As I watched the space shuttle, she seemed like a huge white bird with fire breathing out of her tail. She started to lift up slowly at first, gaining speed and altitude. I would highly encourage anyone who has not seen a rocket launch to go to Kennedy Space Center (KSC). It is, indeed, a sight to see, with a huge white plume against a clear blue sky.

I was filled with pride for my students and pride for our nation. To launch a shuttle into space takes so much effort and sacrifices from so many involved individuals. It was a beautiful and emotional moment. I noticed that others in the crowd were weeping. We were now part of an elite space research family. Something we dreamed about, learned about, and built on the reservation with our mentors was lifting off into space. I thought of our ancestors, like my great-grandmother and great-grandfather who grew up in tipis without electricity or running water. They rode horses and brought their belongings from place to place with travois. They cooked over an open fire. They had no cell phones, TV, cars, or computers. They never even dreamed of an airplane, and in just a couple of generations, their great-great-grandchildren were participating in a project that was being carried by shuttle into space.

I am unsure if the students shared my thoughts concerning their own ancestors and the incredible feat that was happening, but they were elated. We all congratulated ourselves. I made sure the students knew what a job they had all done. Their self-esteem started to fly as high as the rocket that had just lifted off. They all knew that they could achieve a hard project, perhaps like graduating from high school or college (which some did later on) or getting a good job with the tribe and living a good life. It was a proud, good day for all of us. The high bar of research and education that NASA had set was achieved. We made the cut. My students most definitely had the right stuff. I still smile both inside and out when I think about this.

The first space experiment designed and built on a reservation by an all-Indian science team was launched into space on June 2, 1998, at 6:15 p.m. to test a NASA scientific hypothesis: Can a human make fertilizer in space? This represented a huge moment in history, not only for us as an Indian team, but for Indians across the nation. American Indian nations were officially in the space race, led by the Shoshone-Bannock Nation.

STS-91 *Discovery* went on to fly a safe mission, and landed at KSC a week later with no problems. I was invited back to KSC to retrieve Baby. This is called de-integration, when the experiment is off-loaded from the shuttle and taken to the same clean room in which Baby first was integrated. The clean room ensures that there is no contamination of experiments returning from space, and there is no contamination from Earth into space. As I was flying from my classroom to Kennedy Space Center, I wondered what happened to Baby in space. Did she make fertilizer?

A BABY HOMECOMING

When Baby flew on STS-91 *Discovery*, she was in a canister in the back of the orbiter. Once the shuttle was stable and in orbit after liftoff, the astronaut crew would begin their work. Within an hour after liftoff, the crew would be busy with the space mission. One and half hours into the mission, one of the astronauts would flip on a power switch that would power up Baby. Baby would run for thirty minutes and make fertilizer in space. After an hour passed, an astronaut would flip the power switch to the off position. Our experiment would be completed during one hour of a seven-day mission.

During the mission, NASA decided to view the mission in space on its live TV channel. A video camera aboard the spacecraft would be turned on, and the crew's experiments would be followed by observers on Earth. As my students and I watched, the video went dark while the camera showed the payload section. We waited to see that view again. It failed to come back on. I wondered if the power outage would affect our little Baby. I worried about losing power to Baby. The mission continued. No more power issues occurred. STS-91 made it back to Earth safely.

When I was given access to Baby and examined the canister, it looked like nothing had occurred. It was determined by the de-integration team that Baby's power did not turn on for after that section of the shuttle lost power. Baby had gone on a ride in space but had not made fertilizer. This is the way of science. Many times, we learn as much about why things do not work as how and why things do work. As I explained to the students, who were very disappointed, this is life, full of surprises, both happy and sad.

We learned that all the equipment that composed Baby survived the rigors of space. For example, the hospital IV bags that held the water for mixing were exposed to temperatures of minus 280 degrees Fahrenheit, and when the bay of the shuttle turned toward the sun, to temperatures of plus 280 degrees Fahrenheit, they survived. All the lines and the other parts survived as well. What a tough little Baby. But my students were bummed out. This news made them feel like they had failed. I explained that when you fly in space with NASA, there are many pieces that have to work to make a successful mission. This type of learning happened all the time with all kinds of things.

NASA tries very hard to make sure *nothing* fails, because when it does, astronauts can die. After all that work, a power failure in the cargo bay of the shuttle had occurred. I told our team members that not only was our Baby experiment not successful, but neither were any of the other shuttle experiments. NASA would try again, and so would our NASA Club. And there would be more exciting adventures and more experiments yet to come!

CHAPTER 10

Fun with Urine

What is the nature of the universe? What is our place in it and where is it and where do we come from? Why is it the way it is?

—Stephen Hawking (2000, 187)

The rules of the competition were quite clear. If the experiment fails due to the fault of the team, there is no second chance, but if it fails because of NASA, for example a loss of power, then a re-flight consideration is in order. Acceptance is not guaranteed. The case must be presented to the NASA review board all over again, and they get to decide the fate of the experiment. To fly again is a huge honor. It also means that another team does not fly. NASA offered our team another flight on another shuttle. Baby would have to be safety tested again. I thought this was very cool. So we had a NASA students' meeting. What did they want to do? This would be a lot of work to do again and would take a lot of time. The students wanted time to think about it.

We generally have three to six months to design, test, and build the experiment. But our team knew that deadlines could change and delays could be a real possibility. This is challenging for teachers: students get excited and hold on to that excitement for months, and delays are difficult to handle. But that is how real life is; "hurry up and wait" happens with a lot of things in life, including NASA flights to space and back.

Eventually, NASA gave us a launch window. The STS-108 *Endeavour* would carry our Baby up into space sometime between November 26 and December 4, 2001. Eleven students and two mentors would travel to Kennedy Space Center as VIPs and experience the liftoff of our little Baby with *Endeavour*.

But 2001 was a long way off. While we waited, we worked on another project.

SPUDS IN SPACE

Our club discussions eventually led to two additional experiments. The first was growing potatoes in simulated Mars soil and then growing those plants in space (before the movie *The Martian* was released!). My class hypothesized that plants could grow in soil from Mars. Our initial research question was: Could plants grow with simulated Mars soil on Earth?

> The simulated soil, known as JSC Mars-1, closely resembles the red dirt found on the surface of Mars. The coarse powder—about the color of cinnamon—is similar to what scientists know about the color, density, grain size, porosity, chemical composition, mineralogy and magnetic properties of Martian soil (Alexander and Hutchison 2000).

We needed a place to try our hand at growing Idaho seed potatoes in Mars soil. We applied to NASA to get Mars soil and won the proposal to use the simulated Mars soil to grow Idaho potatoes—in the Sho-Ban Nation greenhouse on the reservation, where the tribe was already growing food plants. It worked. If there was enough water, Idaho potatoes could grow with Mars soil. BUT could they grow in space?

The students named the second experiment "Spuds in Space." The research question for this experiment was: Would Idaho potatoes grow with Mars soil in space? Why would we want to know? The students wanted to see if it was possible to make French fries and open a small fast-food palace on planet Mars. I liked their out-of-the-box thinking. Could a rez open up a McDonald's franchise on the moon or Mars? What would be needed? All hands went up to respond: "You need farmers to grow spuds, cooks to cook the spuds in space, and a space/planet marking crew, plus rez astronauts." Man, it would take a whole lot of learning to make this happen! It was way fun to talk and think about the experiment. It was science fiction come to life.

Students mixed wet Idaho spuds with Mars simulated soil and placed the mixture into canisters that would ride in the back of the space shuttle, like a pickup truck in space! There was much excitement to see what would happen.

On May 19, 2000, the Sho-Ban students flew their experiment into space on the STS-101 *Atlantis*. The experiment was highly successful; however, the limiting factor was securing enough water for the potatoes to grow. We think Mars has frozen water on its surface, enough to make plants grow, with the help of greenhouses, of course.

Our story was written up in the *NASA News*. It eventually caught the eye of the Chinese government. They asked if they could do a small story to be featured in one of their life science schoolbooks. Of course, *all* in Chinese. This was so cool to do with the Sho-Ban NASA students. With the population of China at 1.393 billion (World–ometer 2019), we felt that this would be a good story to share with the Chinese people. Other nations beside the United States wanted to know about Indians, space, and Idaho spuds. I had the students answer the questions the best they could while I was the scribe for them. It was a learning experience for the students and I wanted to see

中的土豆

爱达荷州的一名老师。那个州的
加林多先生和他的学生们想知道，
——外太空生长？

天局(NASA)的帮助下，他们做了
种植在由NASA提供的特殊土壤中。
的土壤。“亚特兰蒂斯”号航天飞
空中。在几乎失重的情况下，这些
球上，他们又种植了另一组土豆。
行了比较。猜一猜，他们发现了什
地球上的大1倍！

一样，“太空中的
的问题。为什么
球上长得大
的生长情况
先生和
案。

豆。

爱德·加林多

收集数据

我学到了什么？

1. 加林多先生的学生想知道什么？
A. 土豆能在太空中生长吗？
B. 胡萝卜能在太空中生长吗？
C. 太空中生长的土豆好吃吗？
D. 宇宙飞船上有重力吗？
2. 实验的结果是什么？
A. 土豆不能在太空中生长。
B. 只有一些土豆能在太空中生长。
C. 土豆在太空中长得更大。
D. 土豆在太空中长得更小。

登录 LOG ON 访问www.science.mmhschool.com，学习更多有关太空的知识。

感谢ED Galindo先生授权本书免费使用“太空中的土豆”的所有图片。

The Spuds in Space Story in China.

how they handled the question. They did wonderfully, as usual. The NASA Club had a good time on this project.

I let the students know that this was an example of nation-to-nation goodwill.

SPACE WATER

We previously flew Baby's fertilizer in a microgravity environment on the *Vomit Comet*. Baby produced a fine product. What about something new? Three years had passed since Baby and the original team made the historic *Discovery* flight. I had the next generation of experimenters/students at the school. I formulated an idea for the students to ponder. I would tell the students my story, and then my NASA students would vote.

I called a NASA meeting to talk about what we should do with Baby for the *Endeavor* flight. As I outlined our choices, I asked the students what they wanted to do. We could fly Baby with phosphate fertilizer again or do something else. We had many discussions and questions. We debated the reasons why we would want to fly the same experiment again. I had asked these students what they wanted to do, as they knew it was a huge time and work commitment. Did they want to fly the same things again, since we had a chance to fly again? I asked them, "How about flying urine?" They said "what" kind of loudly. I told them I thought this would be fun. The students thought about this and replied, "Fun with pee?" I replied, "Let's start calling it urine from now on."

The students were learning the functions of the water cycle. Yes, this is in most high school biology textbooks. We recycle water on this planet—and very well, I might add. The water cycle is a beautiful story about how water moves, from the waters of the ocean to clouds over the mountains, forming rain, and returning to the oceans again. Water moves from liquid to vapor to ice and back to water. Clean water is truly a great gift. As with many things that we take for granted, the water cycle has been working here on Mother Earth for billions of years. The question of how water came to be on this world is currently under debate in the scientific community. Some say, "Primordial earth was an incandescent globe made of magma, but all magmas contain water. Water set free by magma began to cool Earth's atmosphere. It could stay on the surface as a liquid rather than evaporate. Volcanic activity kept and still keeps introducing water into the atmosphere, increasing the surface and ground water volume of the earth" (USGS 1995). I have heard others state evidence that while the Earth was being formed, it was being bombarded with comets that contained not only water, but life-giving microbes. Many Indian people view water as medicine, and in some Indian Nations the most scared of ceremonies begin and end with water. Water is a substance that expands when it freezes. This is unique among compounds.

The water cycle is a story not only of water but also of the circle of life. It has no beginning point, and I hope no end! As the water flows over the land, some Indians think of this as the Earth taking a drink herself. Like many Indian people, I believe this. The United States Geological Survey (USGS) calls this "infiltration." Water the Earth uses trickles into aquifers and is stored there. Some of the water in the aquifers locates openings in the land that becomes a spring. "Over time, though, all this water keeps moving, some to re-enter the ocean, where the water cycle ends" (USGS) or begins, depending on where you are in the great cycle of life. Water is a great gift.

I told my students the story of water. We also talked about the chemistry of water, with its bonds of hydrogen and oxygen. Some of the students had performed sacred ceremonies using water. Most knew the value of the gift of water. I told them another story of water. Imagine being on a spacecraft for years. Astronauts cannot carry enough fresh supplies for years of flight. Their survival depends on the ability of the people on board to recycle everything, including using their waste to grow food.

The Sho-Ban Reservation is located in the area of Idaho where there are massive dairy operations. Thousands of cows eat, go through gestation, deliver a calf, and then they produce milk. They also accomplish a lot of other biological processes. They urinate and defecate in large quantities. I imagined piles of manure and golden ponds of urine gleaming in the morning sun. I wondered what happens to all this biomass.

The golden pond of urine gave me an idea. What if we could recycle the water from the urine? This would reduce the pond size and make the urine lake of water less toxic to the groundwater. I thought I would ask the experts I hang around with during most of my waking hours, my science students, for advice.

One morning in first hour chemistry class, I asked how many had urinated before they came to class, and if so, about how much? I asked for a show of hands. You should have seen their faces. I asked this personal question for two main reasons. I wanted my students to think about where water comes from on this planet. And I wanted them to know the water cycle even better than they thought they did.

I asked the students about the idea of flying urine. "Urine," some of the students said. "Why would we want to do that? That is so gross." I smiled, as this is what I do. I love to gross out my science students and make them think outside the box. I explained how I was approached by some Indian members and asked to investigate cultivating food with the mineral zeolite on the reservation. This shows that as a teacher, one should listen to the community you are in. What do they want to do? This makes for great community support!

"What if we could mix urine with zeolites and make another liquid fertilizer?" I asked my students. "Zeolites are minerals that have a small (micro) porous structure," and essentially have the ability to selectively sort molecules by size (Nebergall et al. 1973, 216–217). "You mean you want us to make Baby pee in space?" "Yes," replied one student. "Think of how much fun that would be." They all just stared at me. In the end we voted to make Baby urinate in space. I continued to reassure the students that they would have "fun with urine." I knew very well I would.

I wrote to the NASA flight center and requested a change in our experiment from the phosphate ion and water to "Fun with Urine." NASA officials and a number of academics in universities rejected the name and requested that we not name our experiment "Fun with Urine." I told the students about the concerns of NASA and the university and asked what name they liked. All liked "Fun with Urine." So we kept it, and it stuck.

NASA needed tons of paperwork, documentation, support documents, new methods, analysis changes, chemical changes, mechanical changes, and a broad overview of why we wanted to change our experiment before they would approve it. Of course, this all had to be done yesterday. Did I tell you that teachers do not need sleep for months at a time? Teaching is not for folks who like a nine-to-five job or much sleep, but it sure can be fun, with or without urine.

SIMULATED URINE

> *Native education, at its innermost core, is about life and the nature of the spirit that moves us.*
>
> —Gregory Cajete (1994, 42)

NASA gave us permission to fly human urine with our experiment. My science class knew this. The staff at my school knew this. The parents of my students knew this, and by now the entire reservation community knew this. I am not sure the reservation community or the community outside the reservation knew *why* we were flying urine with NASA. Nevertheless, everyone knew we were flying urine into space, and they were determined to help us.

The opportunity to learn, build, and dream with NASA is a rare honor. From an educational viewpoint, the chance for my students to consider the real-life applications of science and math skills was "way cool" for me as their teacher. From a research perspective, the chemistry needed to separate nitrogen from urine, and capture it, was fun to think about. Getting a new batch of students exposed to engineering, a NASA flight, and mentors was great! These benefits were but a few that I associated with the opportunity to fly with NASA. The community excitement I received to help our NASA Club was a pleasant surprise.

In my mind, schools are reflections of the community's values. Sho-Ban School reflected the Shoshone-Bannock Reservation community. The parent support I had for the project was amazing. Parents came in all the time to talk with me or with the students. Once the story got out to the reservation community that we were flying urine into space with NASA, I started to get a lot more parental involvement.

The announcement went out in the community via newspapers on a Friday. The students and I were excited that folks knew about our opportunity with NASA, and what this meant to the students, school, and Sho-Ban Nation. One early Monday morning as I was preparing for class, I heard knocks on my classroom door. When I opened the door, a small line of parents and community members was forming. I asked, "What is all this about?" They had silly smiles and grins on their faces and said that they had their samples of urine ready for our class to use.

Back in the day before cell phones and the Internet, the moccasin telegraph was a method to circulate information from family to community and beyond. It is still alive to this day. Word spread to the community that the students and I were flying urine into space. The community began collecting the urine for the students' experiment. The urine from the kind, enthusiastic Indian members was presented in all kinds of containers: soda cans, oil cans, milk containers, old water containers, plastic bags, and anything that would hold liquid, sometimes with no lids. Soon I had gallons and gallons of urine to store in my classroom and lab, and more coming in every day. I even had folks from outside the reservation who wanted to donate their urine. I politely told everyone that we had enough urine for now, and thank you. The people were happy to help, and the thought of their urine on a NASA aircraft made them very happy. I was swimming in parental and community involvement!

I called NASA to tell them of the need to change my experimental procedure a bit. They asked for a reason. I said, "How would you like to store fifty gallons of urine in your office for six months?" I also know enough basic human health to know that not all urine is the same. People eat different things, have different diseases, and many of these things show up in urine. We needed standard urine samples to run to make my chemical analysis constant (controls). Could I use simulated urine, if I could make it? The NASA folks replied, "As long as the urea content is the same in each sample and the simulated urine content is close to human values then it is fine." Of course, this had to be in writing . . . tomorrow. I had to figure out how to make simulated urine. More fun learning for my students was about to begin.

One of our mentors worked in the health profession. Sally, the mom of my student Sarah, worked in a laboratory as a medical technologist. Med-techs work with all kinds of bodily fluids, such as blood, saliva, feces, and urine. I asked Sally if she could teach me to make urine. She looked at me and said, "Don't you know how to make urine by now?" "Very funny, Sally," I said. We both laughed and Sally gave me information on the chemistry of simulated urine. It is often used for teaching purposes, and for different standards. I learned how to make it, and I shared this knowledge with some of my chemistry students.

We were now in the business of simulating urine. We could make simulated urine and control the variables of analysis by making the urine formula the same way each time. The students still thought this was strange, but they learned how to do it. For weeks, I would get calls or visits from community folks still wanting to donate a cup, soda can, or milk jug full of urine with no lid. I thanked them politely and said, "No, thanks. We have all the urine we need."

ZEOLITE AND BABY URINE

Zeolites are members of the family of microspore solids known as "molecular sieves." What does this mean? Essentially zeolites have the ability to selectively sort molecules, based primarily on the size of the molecule. They accomplish this because zeolites have pores of different diameters, and each zeolite filters and traps a specific molecule that can fit through its pores.

How does this affect urine? Urine is liquid produced by animals through the kidney, collected in the bladder, and excreted through the urethra. Urine formation helps maintain the balance of electrolytes, minerals, and other substances in the body. For example, excess calcium is normally eliminated through the urine. Urine also excretes ammonia, the buildup of which is harmful to the body and plants. Urine has certain chemical properties. For example, when it leaves the body its pH, acid or base, can be as low as 4.5 (strong base) or as high as 8.2 (strong acid). It also contains urea, a compound that is composed of 75 to 80 percent of the nitrogen in urine.

In agriculture, clinoptilolite, a naturally occurring zeolite, is used in soil treatment. It provides a source of slowly released potassium. When coated with ammonium, zeolite can serve a similar function by slowly releasing nitrogen, and slowly released nitrogen is better for greenhouse and aquaculture plants. Studies (Ramesh and Reddy 2011) have suggested that some crops may be grown in 100 percent zeolite or zeolite mixtures in which the zeolite has been previously loaded or coated with fertilizer and micronutrients. Zeolites also act as water moderators. They can absorb up to 55 percent of water and slowly release it as the plant demands. This property can prevent root rot and moderate drought cycles. It is used for golf courses to maintain the greens.

The Sho-Ban Nation has a zeolite mine. We wanted to use zeolite in our experiment with urine. We added the zeolite to urine, where it captured nitrogen—the 75 to 80 percent found in the urea. After the zeolite absorbs all the nitrogen-rich urea, what is left is what I call "space water," meaning water free of human waste that can be used to water plants or assist astronauts in accomplishing other tasks in space, like painting things. Some elders of the Sho-Ban Nation asked me if I had thought about doing art in space. I said no. They asked why not. I politely informed them that I needed to learn more about making art in space.

Baby was used to create the simulated urine, mix it with the mineral zeolite, and capture the urea, all within a thirty-minute time frame. Baby also had to do this by using its own onboard computer to push and pull the fluid that made the urine pass through the zeolite and make the urine water available to water plants. The content of urea and nitrates could not be harmful for the plants. If this experiment worked, Baby would start with about one liter of urine and give us (urinate) about 750 milliliters or about 75 percent of the original quantity. We had a big advantage here in that we could use the same mechanism we had for Baby's first fertilizer flight. All the programs and mechanisms would work the same; we just substituted zeolite for phosphate, and urine for water.

NASA recycles urine all the time in space, but their process uses it in several forced (urine) columns, repeated several times. It is pure enough to drink. Our little Baby machine could take urine and make it usable for plants and art in a single pass of about thirty minutes. In essence, when our Baby urinates, we make usable space water. "Fun with Urine" was happening.

CHAPTER 11

Space Flight . . . Again

We are continuously engaged in the act of making meaning and creating our world through the unique process of human learning.

—Gregory Cajete (1994, 25)

Students once again held bake sales and fundraisers of all kinds, and had fun. Our fundraising theme became "Help us have fun with urine." We raised the money to return to KSC. It was exciting. When we were ready for the journey to Kennedy Space Center, we were honored with a traditional blessing ceremony complete with prayers, songs, hugs, and good words spoken.

I do not know how many of you reading this story have traveled with a group of teenagers, but they are an experiment in all kinds of human behavior, including laughter, not paying attention to what is needed at hand, and general fun. I had some help from the excellent mentors and teachers who were part of the team. As we were loading on the bus to go, one mentor asked students about identification: "Do you all have a photo ID?" "Yes," they all said, "Let's go." The drive from the reservation to the airport takes three hours. Standing in the airport line, it turned out that one of the students didn't have photo identification. So we had a life problem to solve. How does one get on a plane without photo identification? I asked the student this question, and the answer was, "I don't know." But in response to my hard stare, the student said, "How about a newspaper?" "What do you mean?" I asked. "Well, the newspaper did a story on us. Here I am on the front page with my NASA teammates. My name is on the caption. See?" Sure enough, it was. "Okay, give it a try." He went up to the counter, told his story, showed the gate agent, and the gate agent gave his okay. Nice solution: lucky, but nice. What does this little episode have to do with science? Everything! This is what science is about, solving problems, asking questions (the hypothesis), trying experiments, getting results, and sharing a good story. My student was presented with a problem, thought of a solution, tried it out (tested the hypothesis), and was "rewarded" with an airplane seat. Science is life, life is science.

COCOA BEACH, FLORIDA

Whenever we travelled to Kennedy Space Center we stayed in the small town of Cocoa Beach. It is a nice place, and we found that the people were great as well. We discovered that Cocoa Beach has a long history of engagement with NASA. In fact, my team of students stayed in the same hotel that was first owned by the original seven astronauts in the Mercury space program. I found it amusing that the hotel in which the original seven rested when they were in the "race for space" was now being used by the Native American Science Association (NASA) space team in the "race for urine."

Cocoa Beach was so cool for us. When one lives on a rural reservation encircled by mountains, one does not travel far from home, so the new group of students were excited. Our home in Idaho, at an elevation of over 5,000 feet (1,524 meters), has some very cold winters, with blowing snow and ice. Now, as Indians from the mountains of Idaho, we were in Florida, at a place called Cocoa Beach. Even the name sounded fun. We had rooms off base to stay, and a local restaurant that gave us a NASA club discount for two meals a day. The two other adults and I made lunch for the students each day to stretch out our budget. This worked well. We also had to find a place to do laundry (as we stayed in Florida for about a month).

The NASA students were, at first, a bit overwhelmed by the whole process, in a huge city and new state and (the other) NASA everywhere. We had a team meeting every day after breakfast to plan our day and make sure they all knew where the NASA class was to be held. I told them that for safety's sake, we were going to do everything as a team. No solo acts allowed! So we went everywhere together as a NASA team. It was fun and safe at the same time.

This natural world was strange for my students, with its swaying palm trees and turquoise ocean. Here the earth was flat, like fry bread gone wrong. The other thing that was a bit strange was the high humidity. The air was moist, and warm, even on December first. Generally, unless the students are participating in a sweat lodge ceremony, they never experience that type of moist air. It even smelled different from what we were used to, presenting a strange new learning environment for my team to explore. One thing the students commented on was that "The people here are so nice." I liked to hear this. I just told them to walk in pairs and be safe at all times.

After we had located our hotel rooms, we headed to the beach. One of the Indian members looked out over the ocean for a long time and finally said, "You know, Ed, this is where the mountains are under the water." I thought about what she said from her perspective. That ocean was so calm and flat. The landscape on the reservation was mountainous. She was raised in the mountains and could see them every day. At Cocoa Beach there were no mountains to be seen, and instead of stately green conifer trees, we saw swaying palm trees and a long sandy beach. Such a strange land.

Science and life experiences are truly tied to one another. Each day is a chance to learn. My small "tribe" was learning. Very soon others were going to learn about American Indians.

On the morning of Baby's launch, the sun rose in the east. We saw shimmering heat waves along the horizon. It was 5:30 in the morning. Equipped with a drum and song, the students and some of the Indian member mentors gathered outside the hotel room in a courtyard. Our "Mother Drum," crafted of the finest hide and built by some of my students and their family members, was the central unifying feature of our Honor Song. It was reverently packed for the trip to Cocoa Beach. We watched from the window of the plane as it made the precarious journey up the moving baggage ramp to the cargo hold. Now we placed it in a position of honor at the center of our circle, connected to the earth and soft grass of Florida. We believe that the drum and the drum beat are the heartbeat of Mother Earth, and as we form the circle around her, we are remembering the great circle of life. The main drummer and singer was a good friend of mine, a respected Indian member who was a veteran. He was our cultural teacher as well. Mr. Ernest Whatomy raised his hand. Drum sticks, with heads crafted of brain-tanned, smoked buckskin, were raised in unison, and the song began.

We offered a special morning prayer song. We began singing just as the sun was rising. We sang to honor our special day. We sang to thank the Creator for this day and for others to follow. We sang in respect for Mother Earth and all her inhabitants. We sang loud, from our hearts, and drummed hard with the drumstick. The drum reverberated, and our voices sang our song loud and beautiful. It was the beginning of a good day.

The drumming and prayer song were uplifting. However, some non-Indian guests did not want to be part of the sacred ceremony of welcoming the sun with prayers and drum. In fact, many called the manager to complain, describing our activities using improper English. After we finished our song, the manager called me over for a discussion . . . both he and I smiled and shook hands. I was thinking about Indian education. A prayer song is a great way to greet a special day, and even better if we can teach others an Indian prayer song at 5:30 a.m., like it or not. I also wondered if the original seven astronauts had ever performed a prayer song with a drum at 5:30 in the morning. I bet they would have, if they were asked to do so. They seemed like a group of guys who would try new things.

On the day of the launch, with experimenters, dignitaries, and dreamers from all over the world, we settled into the VIP section to watch Baby fly again into space. My students and I saw people of many nationalities who came to view the space launch. Visitors from Australia, Japan, and New York City were all there to see STS-108 *Endeavour* and crew fly safely. At first my students were a bit shy but with everyone together, people seemed to be more open for chatting. Some asked my students what they were doing there, as we were in the VIP section. I was happy to see that soon my students were engaging in all sorts of conversation with all sorts of people. The shyness left. This was good for me to see. Once we had a liftoff, the students were clapping and cheering with everyone else in our section. At this time we were all one people and cheering for *Endeavor*'s lift off, Once again, the final countdown, 5-4-3-2-1. Liftoff—what a cool sight. Again, we could feel the rocket blast vibrate in our chests before we heard the great boom.

On December 5, 2001, the good crew of STS-108 *Endeavour* carried our little Baby to the stars. The Native American Student Association (NASA) space research team was flying an experiment into space that they, and their mentors, designed and built on the reservation . . . again.

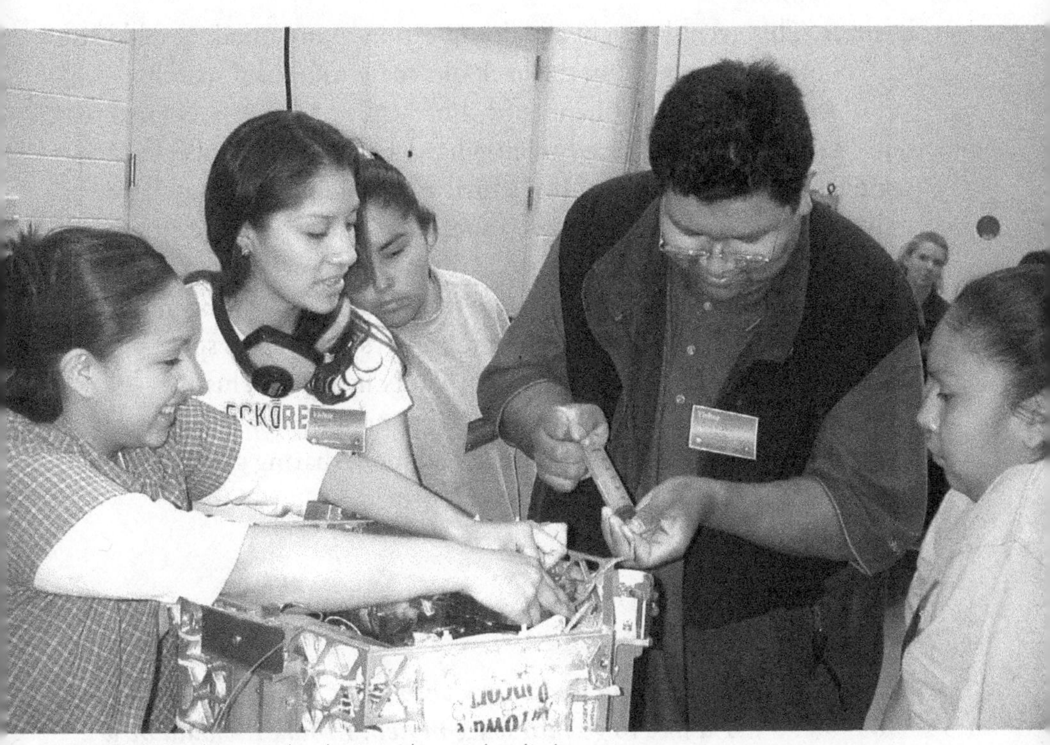

Students loading simulated urine. Photo: Ed Galindo.

Baby flew into space with a liter of our simulated urine. An hour after the crew of STS-108 *Endeavour* was in orbit, an astronaut flipped a power switch and Baby came to life in the payload section of the shuttle. She turned herself on, ran the program, mixed the urine we had made with the zeolite, and made space water. Baby urinated in space, and it worked beautifully.

When Baby came back to earth after twelve days in space, I returned to Kennedy Space Center to retrieve her. Baby had completed her mission and made space water. But something else had happened. She "leaked" into the canister during the mission. Urine water escaped the containment bag. Essentially, the urine mixed with a spore in some excess zeolite, and life began in the rigors of space! This surprised everyone. In the cargo bay of the space shuttle where temperatures reached minus 280 degrees Fahrenheit (minus 173.333 degrees Celsius), with no oxygen, Baby's space water and zeolite had started a reaction. Life found a way to start growing. LIFE found a way to exist in space. This event brought a major question to my mind: Could life have come to this planet on the back of a meteorite from space?

Once I realized what happened, I had called NASA personnel to our clean room and showed them what I had found. I collected the sample with a sterile glove. I had it analyzed in a microbiology lab at a university, and it turned out to be a species of mold spore. Life grew in the harshest environment of space. Baby had completed her mission.

The students and I had the space water analyzed by our industry chemistry mentor, Rob. Rob told us the analysis of Baby's space water showed it was pure enough to water plants in space. This news was celebrated by the mentors, teachers, and community. We were proud of the students. We were all clapping, laughing, sharing gentle pats on the back. We were all having "fun with urine."

THE STUDENTS

> *"I'm proud and happy with what is happening today. I'd like to go into space myself. I think it would be a wonderful adventure."*
>
> —Carmen Edmo, Shoshone-Bannock High School junior and NASA student (Florida newspaper interview the day Baby flew, June 2, 1998)

> *"It's the first time an Indian project has ever been in space. We worked hard, and we're all proud of Baby."*
>
> —Jackie Yokoyama, Sho-Ban junior who worked on the Baby project (*Idaho State Journal* interview, June 3, 1998)

As I reflect on the images of the students who worked on the NASA projects, I think about their motivation and their skills, their good hearts and ideas. Their lives paralleled mine when I was their age in that we were influenced or raised by grandparents in multilingual homes, had a love of science, and worked hard for what we wanted.

I see Angie with her long curly hair and her quick mind. She grasped even the most intricate concepts and shared her ideas with her fellow classmates. Angie would make an amazing neuroscientist.

I remember Shane with his strong opinions about space exploration and how he believed we should explore Earth and the seas first. He worried that our enemies would use space for war. Ella, with her big round eyes and motherly affection for Baby, worried that Baby would be damaged when I took it to JSC without her. Then there was Noah, brought up in the traditional way by his grandparents. Noah, tall and muscular like the warriors of old, had a difficult time blending Western science knowledge with Indian knowledge. Noah eventually understood that both were compatible after learning "two-eyed seeing." Isaack told his classmates, "It should be easy to make fake urine. If you have the recipe, like making fry bread, you can make anything." Maddie, thin and delicate, hesitant to participate, afraid she would do or say the wrong thing. Eventually, Noah helped her to gain confidence just by being at her side. But it was Big Jake who was the team leader. He worked hard to ensure that everyone had a role and that tasks were completed well and on time. Sammy, tall and skinny, was in the after-school science club. He became a leader in the NASA and salmon camps. Sammy is earning a PhD and has become the role model for his children and his wife.

During our NASA Club meetings, the students discussed issues that were important to them. Challenging questions: What do you think of cloning human beings? What is the ethics of going to Mars, when our own planet needs so much help with climate change? Why did the Creator make people racist? Noah was often babysitting his little sister, and occasionally she would ask questions or ask for a story: "Uncle Dr. Ed, tell us the story of how the stars were made, and atoms and how they came to be?" Noah's sister was ten years old and very precocious.

I have lost track of many of these amazing and gifted students. However, when we do meet up on the rez, we always talk about our time at Sho-Ban School. I learned over the years that at least fifteen students have died, but I am unsure how. They seem too young to be dead.

The first time we traveled to Houston to the *Vomit Comet*, these students exhibited bravery in simulated space flight and intelligence far beyond their high school education. They demonstrated that they had the right stuff beyond critical thinking and physical prowess.

Classroom at Sho-Ban High School. Photo: Ed Galindo.

The picture above shows my science classroom at the Sho-Ban School. Observe the round work table surrounded by students working on our experiment. All adults are in the back of the room, including me in the lab coat. Allowing the students to work independently, the adults are standing out of the way. The adults are there as a resource for students, to respond to questions and help. In the photograph, as in the NASA family, students are accomplishing their assignment and learning important STEM concepts. By getting out of the way, allowing the students to work out the problems, the adults are an important part of this learning equation. The philosophies of life and education that come up in the classroom are too numerous to mention.

When students successfully accomplish their tasks, it is impossible for one word or even one sentence to describe what their happiness looks like. On the other hand, it is heartbreaking to see the fire of hope fade from their eyes when their family members or friends abuse them. This is especially difficult for me to accept, even though I know the challenging home environments of some students. Many Indian children are raised by grandparents, uncles, cousins, or some combination of both. Some families are suffering with what I call demons. Demons of alcohol, drugs, abuse, and other life demons. Students may be ashamed to admit that their parents or other relatives are incarcerated, or that they have disappeared. To be clear, not all Indian families have problems, just as not all non-Indian families have these demons—though it appears to me that some of the non-Indian families do have similar torments.

On the other hand, countless Indian families have college graduates, working in professional careers such as nursing and teaching, or lawyers, researchers, and business owners. Numerous family members have progressed from the GED to PhD and are college professors.

We all need hope. Hope for a better day. Hope for better times. Hope for more humor than sadness. Hope for life and not giving up. We hope to keep moving forward no matter our circumstances. For Indian people to live the "Good Red Road of Life" filled with hope, it is important for all of us to be proud of our culture and not feel ashamed of who we are. Our ancestors fought hard so that we could live, be here in the world, and make contributions to our people. We need to honor their sacrifices. Students in the Sho-Ban NASA Club demonstrate these pure qualities. I am thankful for the life's lessons they have taught me. I thank them. They are indeed all "stars." We are all related.

SHOSHONE-BANNOCK
JR./SR. HIGH SCHOOL
FORT HALL, IDAHO

ED GALINDO
Science Teacher
Athletic Trainer

NASA Get Away Special
STS-91
May 1998

Patch designed by Sho-Ban students to commemorate Houston flight.

CHAPTER 12
STS-107 *Columbia*

This story is told in honor and respect of the brave STS-107 *Columbia* crew members—Rick D. Husband, Commander; William C. McCool, Pilot; Michael P. Anderson, Payload Commander; David M. Brown, Mission Specialist 1; Kalpana Chawla, Mission Specialist 2; Laurel Blair Salton Clark, Mission Specialist 4; Ilan Ramon, Payload Specialist 1—who lost their lives in the exploration of space on Feb 1, 2003 (NASA.gov, 2003).

STS-107 was a scientific mission with over eighty scientific experiments from around the world on board, including one from our NASA Club on the Sho-Ban Reservation: "More fun with urine or painting with urine." It was listed in the NASA flight manifest as "Indian Art." The students and I preferred our club's name, and still do!

Due to the success of STS-108 *Endeavour*, our NASA Club applied for this opportunity with the idea of "having more fun with urine." The experiment involved a slow release of urea with zeolite. The hypothesis explored the concept of making space water. The experiment on the *Columbia* mission was twofold. We focused on how various materials would survive the rigors of microgravity. We restricted the materials to wood, plastic, metal, and ceramics that were painted with space water or filtered urine, and mixed with different dyes. The students also wanted to investigate whether painting or art could counteract the effects of space depression on extended missions away from earth, such as a Mars mission. Our NASA Club members interviewed the crew members of STS-107 *Columbia* to ask if art could help relieve stress and what they thought about the "space art." The astronauts were our role models for experiment information. A strong relationship developed among the astronauts and the students. In summary, this was an experiment about the human side of space flight and about being happy in space, at least most of the time.

We used as many different materials as we could find from our school. For example, small wood pieces, metal scraps, plastic, small rocks, rope, and small ceramic pieces (5–10 cm).

All painted objects had to fit inside small self-contained vials that NASA provided for the researchers to use. We replicated our experiments and had three control groups, as instructed by NASA. One control group went to JSC in Houston. The other two control groups stayed in our science lab, one for control and one for backup.

The students filled our vials with the experiment parts and sent them to NASA to be integrated into the STS-107 mission. Once STS-107 was in flight on January 16, 2003, our students would email questions to the flight crew. For example: Did they, and are they, having fun in space? What does fun mean to the astronauts? They also asked questions about the stress of space flight, and if the astronauts thought art could help with that stress. The crew became an important part of the students' lives as they responded and asked questions of the students.

We lost STS-107 *Columbia* on February 1, 2003. We were all devastated to hear this tragic news about our friends. When I learned of the accident, I invited the students and staff who worked on the experiment to meet in our science lab. After such a graphic display of *Columbia*'s destruction on all the news stations, I wanted to be sure that our students and staff were OK. The students came to the lab. Tears were running down their cheeks. Their hearts were broken. The world felt empty. Not only was their experiment destroyed but, more importantly, they lost their friends and colleagues in the NASA family. For the students, the world stood still. We sat in a healing circle and told stories of how we felt. Was there anything I could do to help them through this horrific tragedy?

I let the news sink in for a few days, then called another meeting with the students and staff, again in our science lab. I asked our cultural advisor to be present and say a prayer for our group and the family of the STS-107 *Columbia*. During this time of sorrow, the NASA Club and their advisor decided to conduct a traditional honoring and healing ceremony with the sole purpose of honoring the astronauts, family of the astronauts, and NASA crew of STS-107. Our advisor would instruct the students on what they needed to do. The preparations would take us one year to complete. We would travel back to JSC to pray and sing for the crew and family of the STS-107 *Columbia*. As the faculty advisor of the Sho-Ban school NASA Club, I felt it was important to heal the hearts of the students. They needed this heart-offering ceremony to focus on life and not death, and to help them understand that life and learning do go on, even in disasters. It is an important life lesson for all.

One of our former NASA students worked on the STS-107 *Columbia*'s mission. She was in the rare position of working on the experiment, watching the liftoff in person, and now was on the recovery team in Texas (Chien 2006). In a previous life, she was a member of the NASA Club alumni and helped design and manufacture our experiment. She had become a Sho-Ban firefighter to earn extra money to become a dental hygienist. All my current students knew her and her family.

The Sho-Ban fire department was asked to help with the search, as many Indian fire departments did at this time. The search area was enormous, 2,500 square miles (6,474.97 square kilometers) of forest, swamps, and fields. The Bureau of Land Management put out a call for search volunteers, and firefighters were the first to answer the call. My student got her call in mid-February. She was now part of the search that her experiment was on. She lived in a tent and worked six days a week. She met other crew members from throughout the country. NASA is a true family, and the NASA astronaut family aided in the search and would meet the people searching for debris evidence. One of the people my student met was the first Indian astronaut, Dr. John Herrington. Dr. Herrington was involved in search operations from the command center in Lufkin, Texas, coordinating the search by air (Chien 2006).

The net result was 83,000 pieces from *Columbia*, 40 percent by weight. Each piece was numbered, tagged, and examined. Most of the rest burned up during reentry (Chien 2006). Columbia's debris was stored in the massive Vehicle Assembly Building at Kennedy Space Center. The debris was available for examination to qualified researchers in their quest to understand what had happened to *Columbia*. It was determined that *Columbia* had an in-flight breakup and disintegrated during reentry into the atmosphere on February 1. All seven crew members were killed with the disintegration of *Columbia* (NASA 2020). The ashes, once their earthly bodies, were scattered to the winds.

Students honoring the astronauts on the Columbia. Photo: Ed Galindo.

By examining the damaged pieces and knowing the circumstances of how *Columbia* came apart, engineers could develop better, stronger spacecraft for the future. In the end, *Columbia* helped researchers and engineers make a stronger spacecraft. My student had completed a full circle of life. She was a researcher, excited to see the launch, and was part of the recovery team. She had talked to some of the crew about her class experiment and was now involved in bringing them home.

The following year, the NASA Club students presented the gifts they made and sang their honor songs (led by our Sho-Ban cultural mentor, Mr. Wahtomy) at the Mirror Memorial at Kennedy Space Center. The Mirror Memorial is a massive slab of polished black granite bearing the names of twenty-four astronauts who died in space exploration programs. I was, and I am, still proud of all of the students and mentors who helped to say goodbye to the crew of STS-107 *Columbia*. It was done in honor and respect. During our ceremony, no one had dry eyes.

As a science teacher, one knows that not all science lessons in life come from answering the questions in the back of a science book. Not all science lessons are about science, but can be about the beauty of life itself. This is what teaching science has been for me.

CHAPTER 13

Looking Forward

A long time ago the boundaries between the sky world and the earth world were fluid. Sky people came to earth and animals went to the sky in search of adventure, game or fire. . . . Earthly bears are descended from the celestial bear. . . . Just as the spirit of the celestial bear retreats to its den each winter and reappears in the spring, earthly bears do not die but [look forward to] reappearing each spring after a long winter's sleep.

—Dorcas Miller (1997, 2)

INTERNATIONAL SPACE STATION (ISS)

The paths of education are not linear, but circular. During the time Sho-Ban science class was designing and flying payloads with NASA, our nation's space program took a tragic turn, as STS-107 *Columbia* suffered a tragic accident on reentry from space. As we mourned as a nation, the NASA family researched the cause. Following many months of searching, questioning, and analyzing data, NASA was prepared on July 26, 2005, to attempt another manned flight on the STS-114. I traveled to Houston with two students to experience lift-off and witness a positive historical event.

One of the concepts that emerged from the *Columbia* inquiry was that NASA would fly student payloads once more, but with a different configuration. The new protocol required that all student experimental payloads be brought aboard a space shuttle in the form of a small folding briefcase. The small folding briefcase would passively attach to the wall of the International Space Station—small is better on the ISS. Howell describes the ISS thus:

> The International Space Station (ISS) is a multi-nation construction project that is the largest single structure humans

> ever put into space. Its main construction was completed between 1998 and 2011, although the station continually evolves to include new missions and experiments. It has been continuously occupied since Nov. 2, 2000. As of January 2018, 230 individuals from 18 countries have visited the International Space Station. Top participating countries include the United States (145 people) and Russia (46 people) (Howell 2018).

NASA solicited designs of small experiments that could fit inside the space shuttle briefcase and attach to the wall of the ISS. This was an ingenious idea to trigger excitement for students and teachers about space experiments. The Sho-Ban NASA Club was invited to apply. As usual, I asked the students if they wanted to apply to fly an experiment on the ISS. NASA still has a very competitive process in place. I am glad to say, the Sho-Ban NASA Club won a spot to fly on the ISS. They named their experiment "More Fun with Urine III" (MFWU III). The concept was that students would once again paint objects with rehydrated simulated urine water to answer questions about "space art": Could astronauts paint in space with the club's technique? Would this help the astronauts with their moods, by remembering life on Earth and having fun? Is this something astronauts might want to do on a long-extended mission for many reasons?

The students painted the materials in the classroom, put them in small vials, and sent them to Kennedy Space Center as instructed. From KSC, our little experiment traveled to the Russian space agency in Moscow. From Russia, she flew to the ISS aboard a Russian unmanned spacecraft (M51) on December 20, 2004. The experiment stayed on board the ISS for seven months, then returned to be analyzed by the NASA students. The experiment performed beautifully. During the time the experiment was on board the ISS, students emailed questions to the American astronauts and Russian cosmonauts. The students wanted to know what the crew thought about "fun with urine" and painting. Most of the NASA astronauts thought that not only was art a good idea (some smiled thinking about this idea with urine), but also playing a musical instrument and making music was important for good morale. Learning about the American astronauts' ideas was an exciting surprise for the students and myself. We discovered that

(*Above*) NASA students chemistry/biology science experiment ready to board the International Space Station. Photo: Ed Galindo.

(*Below*) Photo of the International Space Station sent to the NASA students by the astronauts.

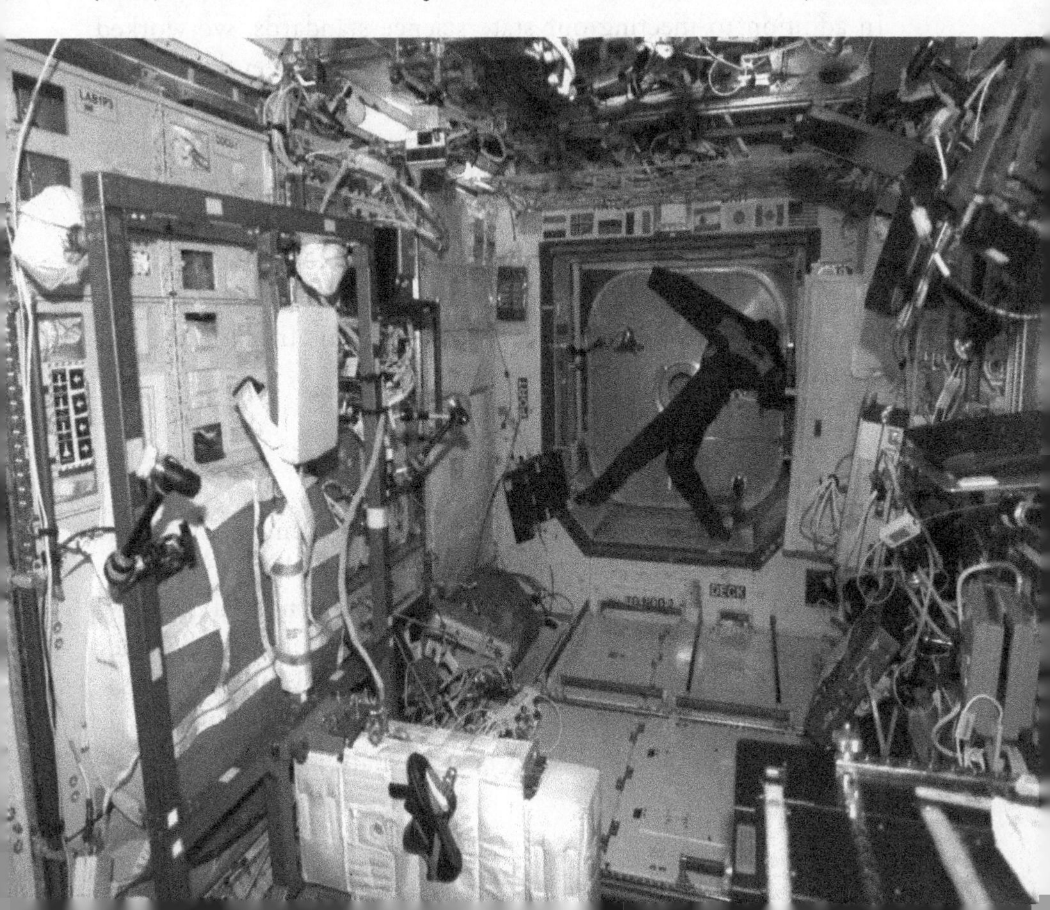

serious-minded Russian cosmonauts would "really not like to be bothered" by student questions. I like to think the cosmonauts actually smiled secretly while having fun with urine, which the students and I found quite hilarious! I mean, who would *not* like to have fun with urine . . . while in space?

PROJECTS HERE ON EARTH

After completing several NASA projects, our team focused on raising fish. We raised salmon and steelhead in old refrigerator boxes that students built in our classroom. We partnered with the Sho-Ban Nation's fisheries department to understand this concept of raising fish in refrigerator boxes. To learn more, we participated in many camping trips to the mountains, four hours away by car from the rez. For most of the summer we ended up camping in the mountains and working with the boxes. Even with community support, this project took a lot of time and effort.

In addition to meeting our state science standards, we worked diligently to have our old refrigerator salmon boxes meet the Department of Fisheries criteria. However, the Indian community drove the students' education and research projects, not the state and federal testing science standards. Currently, I am working with Indian students in the Blanding area of Utah's southwest to explore and fly high-altitude balloons (100,000 feet or 30,480 meters). We are organizing all the standard experiments to record temperature, pressure, GPS, and, with GoPro cameras, looking for plastic trash on the rez. We hope to fly soon.

SUMMER OF INNOVATION

In 2010, I received a NASA grant called "Summer of Innovation." This highly successful project engaged middle school and high school students and teachers on reservations across our nation engaged in STEM cultural activities: building and launching atlatls, playing Indian games such as run and scream and lacrosse, building and launching homemade rockets, measuring their trajectories, timing rocket launches, and getting excited about STEM. A new generation of Indian science

(*Above*) Summer of innovation: Lapwai, Idaho, students ready to shoot off their "rockets." Photo: Dr. Tod Shockey.

(*Below*) Measuring the trajectory of the Atlatl arrow. Photo: Dr. Tod Shockey.

explorers was born. The students were told the stories about the Indian NASA Club and how they were representing the next generation of NASA students. They were participating in the summer program to pay their respects to our nation's shuttle program and to honor the NASA Club they heard about. During the time at the various reservations and Indian colleges, Sho-Ban High School students and faculty encouraged the Indian colleges and high schools to work with NASA.

On July 27, 2011, eighty-five Indian students who took part in the Summer of Innovation stood on the COSWAY at KSC looking at what I thought might be the last American shuttle, STS-135, preparing to fly into space. Two of my original NASA students, who were now part of the alumni group, were with me. All the others were the members of the Summer of Innovation summer school and mentors; about one hundred people in all. I was happy to see such excitement in NASA. As we heard the ear splitting boom when the STS-135 roared to life, the sound once again vibrated in our chest cavities. The bird lifted slowly up as fire burned bright underneath her tail. Students and staff once again dreamed of the stars. I will always keep that memory in my heart, as will all the NASA students. Dream with students, and together we can achieve miracles.

NASA FAMILY AND BEAVER FAMILIES

We shared so much joy and happiness, as well as sadness, that the Sho-Ban NASA Club students and I became family. The students are indeed students, but I consider them family members as well. To develop a familial attitude toward students is common for students and their Indian teachers. I have colleagues who teach at the college level. Their Indian students call them Auntie, Mom, Grandma. This is an important concept and very common among Indians. Many Indians understand the view of family. Let's say the view of life and creation on this planet is connected. Let me give you an example.

My current research project captures live beavers. Students record data on weight, health, etc., and we move them back into the wild to a new pond. Some people have threatened to kill them if we did not move them. The students and I have found that this is a confounding idea for the students and the beavers!

(*Above*) Carrying beaver family to release site. Photo: Ed Galindo.

(*Below*) Elder Lavern Broncho demonstrating how to calm a beaver. Photo: Ed Galindo.

(*Above*) Releasing the beaver family to their new home. Photo: Ed Galindo.
(*Below*) Beaver dam. Photo: Ed Galindo.

Before releasing the beavers, we ponder several questions: Will the beavers like the new home we choose for them? Can we transplant them without hurting them? What do adult beavers weigh? Can we take DNA samples from them? Working with the beaver family, the students and I discovered several things. Sometimes they like their new home, and sometimes they do not. When we transplanted them, we needed to transplant the whole family, or they would not stay.

If one considers beavers as family, related to the human, how would this little water animal be viewed? Would it be trapped and killed for money or to use the smooth pelt to fashion a tall stovepipe Abraham Lincoln hat? Or would the beaver be considered a family member, living in our same environment and constructing its home from sticks and dams, where water creates an improved environment for plants and animals? The life cycle of the beaver is indeed entwined with the lives of humans. An element of the human survival plan is to have an income, but at what cost? Every person has their own response to the question. This is what I ask the students, and I ask you now.

If one is raised to appreciate the ecological principles of the environment, and working and living with Indians as relatives, does that worldview change? This is the hope I have for this planet. I want to view the whole of Mother Earth as our home and relative. Let's take good care of our home and all life who share the planet with us.

I consider my students my relatives. Some are funny. Some have good moods. Some have bad moods. Some laugh at life a lot and make me laugh. Some have miserable experiences in their lives. They are all exceptional gifts who have come into my life. I am humbled and honored to be their teacher. We are all related. This is how I view life on this planet. Many of my students have the same idea as I do. We are all related. It is difficult to express the love I have for them.

CHAPTER 14

Experimental Benefits

> *Native peoples have traditionally applied practical experimentation . . . to find efficient ways to live in their various environments and ingenious and ecologically appropriate technologies were developed.*
>
> —Gregory Cajete (2000, 67)

These are not just stories about what a rez teacher can do with NASA, but more importantly, what ANY teacher can do with student learning outside the state curriculum if they want to. These are lifelong lessons to teach ANY student. Sharing fun and laughter in life while students learn is a good gift to share. We can also learn in many different places in addition to our little classroom; for example, the International Space Station, a science and engineering laboratory in orbit around our planet. This is *soooo* cool! Science knowledge lives in many places, and not all knowledge is found in the back of the science books by answering questions. The students were shown that they need to *ask* their own questions, and discover the answers to their own questions. I asked the students to take responsibility for their own learning, a lifelong lesson. Our dreams continue and I still have very high expectations for *all* students that I work with. We are all "Children of the Stars."

What did we learn from working with NASA? Perhaps this a better way to put it: What is NASA still learning from a group of highly motivated Indians from Fort Hall, Idaho? As a member of a minority group and a science teacher working on the Shoshone-Bannock Indian Reservation, I saw firsthand the boundless potential of the Indian students. I was raised with the knowledge that we all possess many gifts. For example, each day I saw the gifts of humor, art, science, math, history, storytelling, and culture that my students exhibited. However, I am unsure that all students or people see or know about the great gifts that Sho-Ban students possess. I was taught that the gifts one receives from the Creator should be used in a good way to help people. All students have gifts. All students are important. When a student quits school, or the school quits on them, the student may not even recognize the gift that they have given up. Then students may develop the "habit of quitting." For example, when some students drop out of school, they sometimes quit trying to become the best they can be. Getting a job without an education is hard, and the job is boring sometimes, so young people quit, and life becomes difficult and unpleasant, and they stop trying. In the extreme and very sad cases, they quit by taking their life. We all lose when this happens.

In his book *Teaching the Indian* (1999), Hap Gilliland states that there is no single easy reason why Indian students drop out of school. I could not agree more. Students quit for many reasons: school curricula, poverty, racism, lack of hope in their community/Indian Nation, and the student's own attitude. My goal in teaching is to help my students understand not only the basics of science, but also why science is so important in their lives. A problem-solving, hands-on, heart-on approach to science—like "Fun with Urine"—emphasizes scientific thinking and teamwork skills over the simple memorization of facts or questions found in the back of many science textbooks.

As a science teacher/researcher, I define success as seeing students working and playing together as a team to help each other find solutions to all varieties of problems, both academic and life problems. Our work with NASA provided the perfect environment for building student confidence in their ability to solve both science and life problems. It showed the students that even a small reservation school can compete with big schools. By not giving up on very challenging problems, like in our creation story, and strategizing ways to make a space flight deadline on a budget, my students learned that they could compete with the best in the world. It also confirmed the story that Indians students, like their ancestors before them, could solve difficult problems and have success as a family, team, or Indian Nation. It is a great honor to share our story.

The Indian views of natural curiosity, problem-solving, information gathering, public speaking (which is very hard for all students), persistence, and patience were all used to meet the high expectations that NASA developed to fly an experiment in space. My students learned about all these skills with Baby.

Students also learned about long-term goal setting. With NASA one does not get space flight approval for experiments and fly the next day. My students learned to have patience with a long, carefully thought-out process, with delays. The process takes a very long time because human lives depend on the safety and accuracy of a mission. Long-term planning with young students is difficult because they tend to think in such short-term ways. For example, for years, I have been telling students that they need to prepare themselves to go college by taking as many math and science classes as possible. Many students do not see this as an important thing to do immediately. If NASA projects can show the students that NASA prepares for missions in a long-term way, in the same way that sports figures practice long hours to prepare for competition, then students can begin to see the importance of preparing for long-term goals in science and in life.

I appreciate John Dewey's words about education: "Only by being true to the full growth of all the individuals who make it up, can society by any chance be true to itself" (1899, 15–16). All students have gifts. All students are important. What we do with those gifts—now that is another story.

CHAPTER 15
Final Thoughts

> *Two roads diverged in a wood, and I—*
> *I took the one less traveled by,*
> *And that has made all the difference*
>
> —Robert Frost (2002)

This is an optimistic story about the smiles, good hearts, and positive actions that students and I accomplished while I lived and worked on the Sho-Ban Reservation. The reservation is truly a beautiful place, with landscapes that are breathtaking, amazing natural resources, and beautiful people. The reservation is home to some of the most fantastic and funniest people living on the planet.

When stories are told about the reservation, or reservations in general, I am not sure that the beauty of the people, their sense of humor, or the splendor of incredible land is expressed. So many stories from reservations paint a negative picture of despair, drugs, alcohol, poverty, or sorrow. These things, as in any community, do exist, but the Sho-Ban people have so much to offer.

SACRED LEARNING PLACES AND MESSAGES OF HOPE

I like to think of schools as sacred learning places, because they are to me. They are places where we can discover a myriad of things. They are places of freedom, the freedom to think and explore and to ask a variety of questions. So you see that schools are special and, in my case, my classroom had been blessed by an Indian medicine person. I treated my classroom and the people in the class with respect. However, sacred learning places are not only classrooms with four walls and a ceiling. They are everywhere, in the mountains or meadows or in ourselves and in how we treat others.

Why have sacred learning places? Because as Morris Manyfingers from the Alberta Learning Center told me in a conversation in 2004, "Teachers don't teach subjects, they teach people. To teach people, they must make an honest effort to get to know them, spend time with them, care about them, and believe in them."

The NASA students and I built a respectful relationship. I trusted and believed in them. To this day, I truly care about them and always will. We will always be a family. Once this heartful relationship is built, the students know I will do anything I can for them, and they will do anything they can for me. Like build a Baby, fly it in space, and have fun with urine.

Believing in all students, but especially the Indian students, is crucial for the students' success. The disturbing educational success rates for Indian students in comparison with their peers have been well documented for many years, with dropout rates reported for some tribes of excess of 40 percent (Indian Report, 2003; Camera 2015). The story I share with you is not about failure but success, the success of helping students dream that they can compete with university students and accomplish the impossible dream. It is a story about an educational approach to STEM problems that does not just answer the questions in the back of a science textbook. This is a story about creating some positive solutions for students with "hearts on, heads on, and hands on"—real solutions to real problems.

We are at a time when teachers are demanding more and more from their students. Students, teachers, schools, and parents have but a single goal: to find success in education. To have the student successfully pass performance exams is but one component for students to find success. Compassion in teaching and learning from one another is an important part of education. Compassionate teaching is not listed on any education curriculum that I have seen, but it should be. As a country, sometimes we forget about compassion. This story is an attempt to help the teacher, especially teachers of Americans Indians, to ensure that their students not only do well on the assessments, but hopefully be engaged in lifelong learning opportunities. When we truly believe in them and ourselves, *all* students will do remarkable things. Schools are involved with various reform activities and testing. Most teachers are remarkable. They teach because they love it. It is hard work and those who teach know this. I have heard folks say, "If you do not find success in *xyz*, then you can always just teach." What a dumb statement.

Indian students need to be reminded that Indian peoples have always been scientists. Our people discovered that beans, squash, and corn grow better when planted together. The corn removed the nitrogen from the soil and the beans replaced it. The leaves of the squash covered the soil to reduce evaporation. Indian scientists discovered that smoked buckskin is waterproof and unsmoked is not. They learned about the medicine plants through trial and error and by watching animals choose what plants they ate to treat certain ailments. Indian people discovered that the willow plant was medicine to them and reduces fevers (now we know it contains an aspirin compound) and that the yew plant or rose hips prevent scurvy. Today, we call this Indian science. It is holistic knowledge of the environment learned by a people who have lived in a certain place for hundreds, maybe thousands, of years. Our ancestors may not have had the word "science" in their language, but they were operating on the scientific method of making a hunch or hypothesis, observing what happens or testing the hypothesis, and repeating the observations. "For our early Indian scientists, maintaining their relationship with nature was critical to survival, both physically and spiritually" (Fox and Lafontaine 1999, 193–201). The knowledge held by the Indian scientists was phenomenal, and is still today.

MENTORS

A mentor is a person or friend who guides a less experienced person by building trust and modeling positive behaviors. There is a misunderstanding about what it takes to truly educate and mentor our young people. As a Yaqui Indian teaching on the Sho-Ban Reservation, I know that true education cannot be accomplished by teachers alone or by a school system alone, nor should it. True education can only occur with the caring help of a considerable collective of people such as an Indian Nation and mentors, community, industry, university, and government programs that serve the needs of the community. On a reservation, an educated person with knowledge may not look the same for everyone. For example, an Indian community may follow the concepts of Indian science rather than the beliefs of Western science, but they know learning is important.

At times teaching can be quite lonely. It may seem odd to a non-teacher that when one is surrounded by students all day long, a teacher can feel alone. When the bell rings for class to start and the door is shut, there you are as a teacher, standing before a class of many young, hungry eyes. Teachers need help. Many do not have enough classroom supplies. Some students need more help than others. Basic everyday help is truly where mentors shine. Mentors need to be brought in to work with students and tutor in the areas that the head teacher suggests.

Mentors can be the parents of the students, community members, alumni, university folks, industry, and many others, all with a shared goal to support the teacher to reach the students. In our story, mentors were so very cool, and they gave so much of what I like to call "heart time." They cared about my students and the community. Each worked incredibly hard with me. They know who they are in this story.

I developed an alumni group for my NASA experiments so that students who graduated could still be a part of the team, because they initially worked on the experiment and wanted it to fly. Some members of the alumni group were college students or they were working for the Sho- Ban Nation. This was good for current students to see and understand that commitments take time, but they are still important to make. It gave older students a chance to become mentors, to help the younger ones learn and become excited about learning.

COMMUNITY

The community of any school is extremely important, but for Indian education it is even more important. This takes hard work because some in any community see the school and the students as separate entities from the community. They are one and the same. The school is an extension of the community. Teachers should learn to include the community by inviting the parents and grandparents into the classroom, not just for a parent-teacher conference, but for a purpose, so they can see their students shine, and not just their own student, but all students in their community. Parents are proud of their students. Teachers must give their students the opportunity to shine for their community members many times a year. Have plenty of good snacks for everyone.

LOCAL BUSINESSES

Other, often unacknowledged, partners are local businesses. Many times they have what schools need: time, funds, tools, and often an internship opportunity. The local businesses in this story were super good; they just needed to be invited. We obtained some of our actuator parts at the local hobby store. Businesses can find specific ways to help their students. Often, they are more willing to give discounts on products rather than donating money directly to a class. I would have had a difficult time without the help our local business community offered me and my students.

Another interesting fact about having a local business mentor is that they can provide an internship for students. Many times, internship opportunities for students are just as important for the team as anything a teacher can do. Businesses are just waiting to be asked.

NASA

To be an effective program, teachers ought to demonstrate to students how their education dovetails with their community, important organizations, institutions, and businesses within the larger world. Aim high! Connect students with enterprises that promote knowledge and have a high bar for success, like NASA. By offering me a NASA fellowship to complete my PhD, NASA set me on a challenging journey. My program was demanding and NASA forced me to think, but it was fun. They set high expectations for me, just as they had high expectations for my class experiments to fly in space. I liked to tease the NASA professionals. Even today, I still like to kid and joke with them. I like to ask them if we can have fun with urine, or poke at them about diversity. They are improving their attitude about my jokes and humor. I encourage NASA to continue the student and teacher programs. This is so important for our students to dream about. I think NASA doesn't realize how much influence they have on the hearts and minds of students and teachers in this country. I know this from firsthand experience. We all need to dream and aim high.

UNIVERSITIES

The university community is a gold mine of ideas and resources. I know university professors and administrators want to help, and there are specific things they can do to help students. They have "stuff," like lab materials, and students to help with the school's science labs, as well as opportunities for field trip learning. They are a natural next step for the students. Universities make good partners, and the professors care about the students. A key step for the high school teacher is to create a positive relationship between the university community and the school community. The university community also offers one of the most important items any mentor can offer: hope. They offer a hope, a pathway to education, and perhaps a lifelong interest in learning. And so, I hope one begins to see that education takes in its many forms with the input of many people. In the end, the students benefit.

FAMILY

The teachers in a small reservation school are family. Like all family members, some are more fun than others. However, all are important. The single most important thing a teacher can do for Indian students is truly get to know them and believe in them. Set the bar of expectations high. It is critical for Indian students to know that teachers care about and believe in them. This is essential for all students, but is especially true for Indian students. This is important because after a while, a good teacher is considered an extended family member, and as an extended family member the teacher is expected to treat students in a respectful way. In turn, the students reciprocate and treat the teacher as a respectful family member. A positive teacher can have tremendous positive effects on the lives of the students. However, a negative, mean-spirited teacher, regardless of skin color or tribe, can negatively affect a student's self-esteem, and rebuilding that self-esteem may take a very long time. My advice here, if you are not 100 percent committed to the students, Indian or not, *get out of teaching.* Not everyone should or can be a teacher.

STUDENTS

Finally, I want to talk about the Indian students themselves. Many live in challenging situations. In spite of their life's difficulties, their hearts are moral, and the students are hopeful for a better future. I have seen their smiles and heard their laughter when at times they did not have a lot to be happy about. They found their way. I can relate very well. They are my heroes and sheroes. They have struggles like all people, and are tested like we all are, but I would encourage all students (Indian and non-Indian) not to give up on their dreams. I do not have any doubts that the students can be successful and give back to the community. I have seen what Indian students and a community with dreams, determination, and good hearts can accomplish. This is what is important.

One of the greatest needs of Indian students today is self-esteem. Students need to develop positive and wholesome feelings about who they are and what incredible gifts they have to share as Indian people.

Sho-Ban students in traditional women's dance regalia. Photo: Ed Galindo.

Indian students are as bright and exceptional as any other students on this planet. They should be proud of who they are, hold their heads and hearts high. I am proud of all my students.

Since this story occurred, some of my students now walk and live among the stars—with the Creator. They won't be forgotten nor the work we accomplished as a NASA team with a little Baby. We are truly one, and truly we are all related. All my relations. This is their story.

CONCLUDING THOUGHTS AND YOUR QUIZ

You have not yet finished! Much like my Grandmother did, it is time for me to give you a little quiz.

You were given many, many stories to think about. In fact, if you really think about what you have read and learned, you should see that there were stories within stories. This is how I was taught to listen and think. Did you catch them all? Let us check a few things.

One of the most important ones was an example of the "Four Rs." The four Rs I am talking about are Respect, Relevance, Reciprocity, Responsibility (Kirkness and Barnhardt 2001; Kovach 2009; Wilson 2008).

Respect: Respect refers to the cultural integrity of Indian people. If you closely examine the story about "fun with urine," you will find that the Sho-Ban Elders asked about bringing art into space. As I reflected on how art makes meaning for me, I found the concept of art in space was also an easy way for the NASA Club students to understand the concerns of Elders and their questions. Our club respected and was motivated by the Elders' advice. The students' knowledge of their cultural art and the advice of elders was respected and valued.

Relevance: Relevance indicates that the work dovetails with cultural perspective, and contributes to community survival. All the research is community specific. Again, the "fun with urine" story drew on Indian knowledge of art in asking the question, why is art so important to take into space? Since time immemorial our elders thought of the next seven generations. They left messages in their art and in stone petroglyphs for us to learn their meanings. Messages left by the ancient ones on rock that we see today are both art and messages to be interpreted by the reader. The meanings may be different for each reader. Someday Sho-Ban art may be left on Mars for others to interpret.

Reciprocity: Reciprocity involves giving back to the community, which contributes to community empowerment and self-determination. Our work with the Sho-Ban students and NASA infused empowerment and notoriety into the community. No other Indian tribe had attempted to accomplish what Sho-Ban high school science students achieved. It was not through basketball or other sports as is the case on other Indian reservations, but through science. Reciprocal relationships also involve giving back to the students. Students know that the teacher is not always the "creator and dispenser of knowledge." Fun with urine was only one way to think about life and science. The Sho-Ban students had their input as well.

Responsibility: Responsibility indicates that as we travel, complete our research and our projects, we consider the responsibility of our spiritual beliefs and our culture. An example of this responsibility is the 5 a.m. drum ceremony in Cocoa Beach.

As Kirkness and Barnhardt (2001) stated, students are seeking an education that respects them for who they are, is relevant to their view of the world, offers reciprocity in their relationships with others, and helps them exercise responsibility over their own lives. This is an education for the students and teacher.

As we reflect back on our story, do you follow all the points? Do you understand Elder Marshall's "two-eyed seeing"? It is a powerful but simple concept to use.

In conclusion, we circle back to the creation story that was told in the beginning of our story. Did you get the point that this is symbolic of what the Shoshone-Bannock students encountered with the NASA problems they faced? Did they keep focused on the problem of getting a scientific payload into space to orbit the earth without getting "distracted" by life as they made their way? They did a magnificent job of thinking and doing great science, measured by NASA.

How did you do as a reader? Did you pass your quiz? If not, read and think about it again!

DISCUSSION IDEAS: CEREMONY IS EDUCATION

Do you want to know why we do a morning song and prayer?
Do you know why we drum?
Why do we hug one another before school?
Education Is Ceremony.
We acknowledge education and the creator.
This rock, this eagle, this feather, this tobacco
all these things are natural.
We are all connected. These are all our relations. We are all living and human
[Teachings of Elder Francis Whiskeyjack, March 11, 2011]

—Simmee Chung (2018)

EVALUATION

Outsiders might ask: How did the Sho-Ban science students do? What did they learn? Each spring, all high school students in Idaho, including the Sho-Ban students, are required to endure the Idaho State Science Assessment test. The years we were flying with NASA were no exception. The students are responsible for all the science sections of the state assessment: physics, biology, technology, engineering. They are also tested by the BIA under the Indian Education Improvement Act of 2001. (This act was the beginning of the BIA assessments for school accreditation.)

In addition to these grueling assessments, the Sho-Ban school board has their own criteria for the academic success of their students. The Sho-Ban school board, like all school boards, wants the best for their students. The Sho-Ban school board wants to guarantee that their students learn about and remember their heritage and the Indian Nation's treaty. They want their students to understand what it means to be a sovereign nation. They want their students to know how easily the Sho-Ban's treaty could be broken by the US government. It has happened before.

A few members of the tribal council indicated that they hoped the students learned to grow potatoes. Why would this be an issue?

Because of the approximately 521,519 acres of reservation land, over 95 percent was leased to non-Indians. Indian Council members hope that eventually the students will assume the agriculture business of the Sho-Ban.

These assessments were in addition to the endless list of NASA requirements concerned with flight and human safety. The students worked hard and they passed *all* the tests. One may ask, "Was this easy?" *No*. Sometimes they wanted to surrender. But something marvelous happened. They formed their own science and math study groups called "science family attitude." Not all students attended the study group sessions, but the study groups demonstrated how much they cared for one another and how concerned they felt for each other. When we were in Texas at Johnson Space Center, their bonds grew much stronger. They felt that they only had one another for support in passing all their exams. The students continued to demonstrate this caring behavior even when they returned home to the reservation. Students who were not on the NASA team inquired whether their NASA team classmates were faring well.

Another question outsiders might ask is: "Was the NASA experience good for our students?" This is a fair question. I would answer yes, for many reasons. Our students had opportunities to test themselves with the best and brightest across the nation. Our students found self-esteem and success within themselves, not only with a successful Baby experiment launched into space, but by passing the demanding NASA physiology and flight tests required to fly. The education bar was set very high by NASA and was not lowered because they were Indian. It was huge for the students to realize that they were the only group that did not vomit on the *Vomit Comet* ride. This in itself was a life lesson. Just as the story in the beginning of this narrative, students had an assignment, and they passed the test and completed the task beautifully. NASA set a deadline and the students met that deadline. No special consideration was given to them. They just did the required work. They learned that they can accomplish countless tasks, if given a chance. They learned about fantastic places outside of the reservation. The world is indeed is a wonderful place if they work to make it so. They made friends outside the reservation. We all had fun learning.

COST

The lesson for school boards and educators is that it does *not* take a lot of money to learn cool things with students. Students and teachers can raise money and ask for support for things they need, if they all work together. It is best to have a very cool idea first, then find the money to make it so!

Another possible cost is time. I believe that the time spent learning outside the classroom was well worth the effort. Students can learn everywhere, in school or out of school. The key ingredients are a creative, dedicated teacher and a student who wants to learn. Every student has a unique learning style. A teacher needs to be creative enough to teach to every learning style, whether it be hands-on, lecture, or kinesthetic.

REPLICATION

Can the Sho-Ban experiments be replicated in other classrooms? I would say, yes! Teachers do not need to replicate a NASA experiment, but they do need to have an interest in their community needs: for example, an interest in improving water quality or restoring salmon runs. When teachers are in tune with the community, then what to study outside of school is not difficult to figure out. It is important to engage the community and parents in the education process. To educate our youth, it truly takes a community/village, not just a teacher's point of view, standard education curriculum, or state standards. Back in the days before contact, this is how Indian people educated the youth. We learned from the natural world, our environment, and the experiences of our village.

STUDENT ENGAGEMENT

For the teacher to engage students *all* the time is not an easy task. I found that to have a science classroom goal helps. For example, I had a goal of increasing the rate of salmon numbers in the wild or flying an experiment on the International Space Station (ISS). These are extraordinary examples of classroom activities that engaged my students and the community as well. They created a spark of excitement.

It is then up to the teacher to make the fire. Not every teacher can have as lofty a goal as flying with NASA, but all can engage the students and community in what they want to achieve. For example, a community might be interested in providing habitat for butterflies or ensuring that turtles and frogs have safe passage across a busy highway.

THE FUN IN LEARNING

I am not sure when some teachers and students lost the ability to just have fun while students learn. Learning takes discipline, but to not enjoy or have some good fun learning and teaching makes me wonder, why do it in the first place? If it is just a job to learn or teach, then I feel something gets lost in the process. If teachers find the material boring and not fun to do, then I would bet students feel this as well. Find some way to make what is being taught fun and interesting for students. The results are well worth the effort.

Who is Lori?

This is truly Ed's story, and it humbles me to be partners with him in this project. I have known Ed for at least twenty years. His heart is as immense as the universe. He loves his students, their families, and teaching. I call him *Superman* because he does such heartfelt and amazing work with Indian students. There are no words in any language to describe the incredible work he does for elders, children, families, and the animals in the natural world.

My culture is from French Acadia and the Maritime provinces of Canada's east coast and New England. As a Nulhegan Abenaki and Mi'kmaq woman, I was taught to use my education to make peoples' lives better. Like Ed, I also grew up in a four-generation, multicultural, multi-language, English-French home in New England. As a family we took frequent trips to visit relatives and the shrine of Ste. Anne de Beaupre in Quebec, Canada. My memere gave me my middle name Anne, who is the patron of the Mi'kmaq people. My grandfather gave me my Indian name Sismoqn Kiwnik (Sugar Otter). We are from the Lentuk (Deer) Clan. My mother named me Lorelei after the Lorelei Rock in Germany, so named for a German legend where a beautiful maiden threw herself into the Rhine River in despair over a faithless lover. She was transformed into a mermaid or siren who lured fishermen to destruction with her beautiful singing. Mom told me I was the only Lorelei in the world. After defending my unusual "pagan" name for twelve years in Catholic school, I followed the desires of my grandmother and enrolled in Cambridge Nursing School.

After passing the Registered Nurse examination for the Massachusetts State Nursing Board, I moved for two years to the Philippines and worked with Maryknoll Missionaries in Manapla, Bacolod, at the Victorias Milling sugar plantation's Saint Joseph Hospital. When I returned to the states, nursing practice was very different from my work in the Philippines. It entailed more paperwork, more theory, and fewer face-to-face encounters with patients. Disillusioned, I returned to school for a bachelor of science degree in therapeutic recreation at Temple University in Philadelphia, and graduated *summa cum laude*. Working with in-patient adolescents as a therapist, I became a believer in outdoor recreation therapy. The natural world is effective at healing stress and anxiety. This prompted me to seek a master's degree in environmental science at Acadia University, also in Philadelphia. I graduated *summa cum laude*. I secured a position with the Schuylkill Center for Environmental Education, where for fourteen years I taught environmental science courses for the Pennsylvania State University, Temple University, and Beaver College. As the resident naturalist and environmental educator, I led members of the center on eco-journeys to observe humpback whales on Georges Bank, a large elevated area of the seafloor between Cape Cod, Massachusetts, and Cape Sable Island, Nova Scotia, Canada. I led trips to view migrating waterfowl in Brigantine, New Jersey, and to experience the birthing of harp seals on the Magdaleine Islands, a small archipelago in the Gulf of Saint Lawrence off northern Quebec. On my first Arctic eco-journey to study Arctic ecosystems and polar bears, I was inspired to return to school once more for a PhD in medical ecology with a focus on Arctic studies. Medical ecologists research the effects of contamination on human health. I viewed the planet as my patient.

It was in the Arctic that I met my future husband, a filmmaker and broadcast engineer from Montana. After finishing my coursework at the Union Institute and University in Cincinnati, Ohio, and a summer internship at the National Science Foundation (NSF) site in Toolik Lake, Alaska, I moved with him to the stunning mountains of the Flathead Indian Reservation in Montana to write my dissertation. The Nursing Department at Salish Kootenai College (SKC), a tribal college on the reservation, was seeking a staff member to work with a Kellogg Grant for distance education and I was hired. Of course, Montana has no Arctic, but I had experience with learning online with my PhD. My supervisor offered me the opportunity to go for a postdoc in education and technology at the University of British Columbia, Vancouver, Canada. At that time, the college administrators were motivated to initiate a distance education program for the entire school. I was hired for multiple positions: curriculum coordinator, Title III director, later the director of distance education. Working at SKC was an amazing experience. I loved teaching the students. I loved showing them how important they are. I loved watching their journey from insecure students to students with high self-esteem. With my encouragement, many of my students finished graduate school when they didn't believe they could succeed!

In 2009, my colleague at SKC, Dr. Carol Baldwin, invited me to develop a course in Indigenous research for her class of psychology students. It was remarkably successful. I began to question how other Indigenous groups viewed research and researchers. In America there is a dark history of researchers entering Indian communities to conduct research without permission and doing damage to Indian people. To find the answer to my question, I was awarded a faculty research scholarship from the American Indian College Fund to explore how other Indigenous peoples viewed research and researchers. Carol and I traveled to Australia, Canada, and the Flathead Reservation to interview Indian tribal members and Indigenous peoples. During our Australia trip, I was introduced to the Australian Indigenous Research Association. For our Indigenous peoples here in Americas, we needed a similar association. So, the American Indigenous Research Association (AIRA) was born here in America. The association has over fifteen hundred members, and every year holds a conference so Indian faculty and students can demonstrate their research projects. Ed is a strong supporter of AIRA. We met as a result of my scholarship with the American Indian College Fund, where Ed was a board member. After twenty-two years at SKC, I accepted a position as a community research associate with Montana State University in Bozeman.

Over the course of my career, I have written six books, presented as an invited keynote speaker in Finland, France, Canada, Australia, Norway, and national conferences, and raised three children and four grandchildren. Being a grandmother gives me the greatest joy! I continue to live on the Flathead Indian Reservation with my husband, my two daughters and their families, and the spirits of my nine huskies. My daughter Emily is the director of the county health department and my daughter Gina is a social worker in Missoula. My son, Ted, lives in Philadelphia where he is the director of Carelinks, an agency assisting in the transitions of in-patient individuals to society.

Wala'lioq for listening.

Dr. Lorelei (Lori) Anne Lambert, PhD

APPENDIX 1

Total missions the NASA Club "flew"

KC-135A *Vomit Comet*, May 15, 1997

STS-91 Shuttle *Discovery*, June 2, 1998
"Fertilizer in Space"

STS-101 Shuttle *Atlantis*, May 19, 2000
"Spuds in Space"

STS-108 Shuttle *Endeavour*, December 5, 2001
"Fun with Urine I"

STS-107 Shuttle *Columbia*, January 16, 2003
"More Fun with Urine II"
(shuttle and crew lost)

Progress M51, Russian unmanned rocket to International Space Station, December 20, 2004
"Even more fun with Urine III."
Stayed on board for seven months and returned with STS-114A, "Return to flight," launched July 26, 2005, landed August 9, 2005, Edwards Air Force Base, California.

APPENDIX 2

Schematics

Phosphate Ore Cylinder Assembly

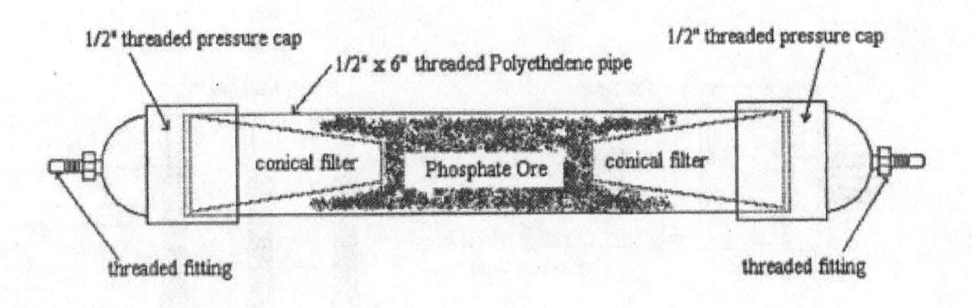

Material Sources:

1. threaded Polyethelene pipe { and conical filters (or similar) }

Orbit
845 North Overland Road
North Salt Lake City, UT 84054
(800) 887-8873

2. threaded fittings:

Antelco
(800) 869-7597

3. threaded pressure cap

Agrifin Irrigation Products, Inc.
1279 W. Moraga
Fresno, CA 93711
(209) 431-2003

Other materials not part of assembly:

4. T - fittings (p/n T250) and elbow fittings (p/n EL250) :

Agrifin Irrigation Products, Inc.
1279 W. Moraga
Fresno, CA 93711
(209) 431-2003

5. Nylon check valve:

Fourmost Products
4040 24th Avenue
Forest Grove, Oregon 97116
(503) 357-2732

G90 System Diagram

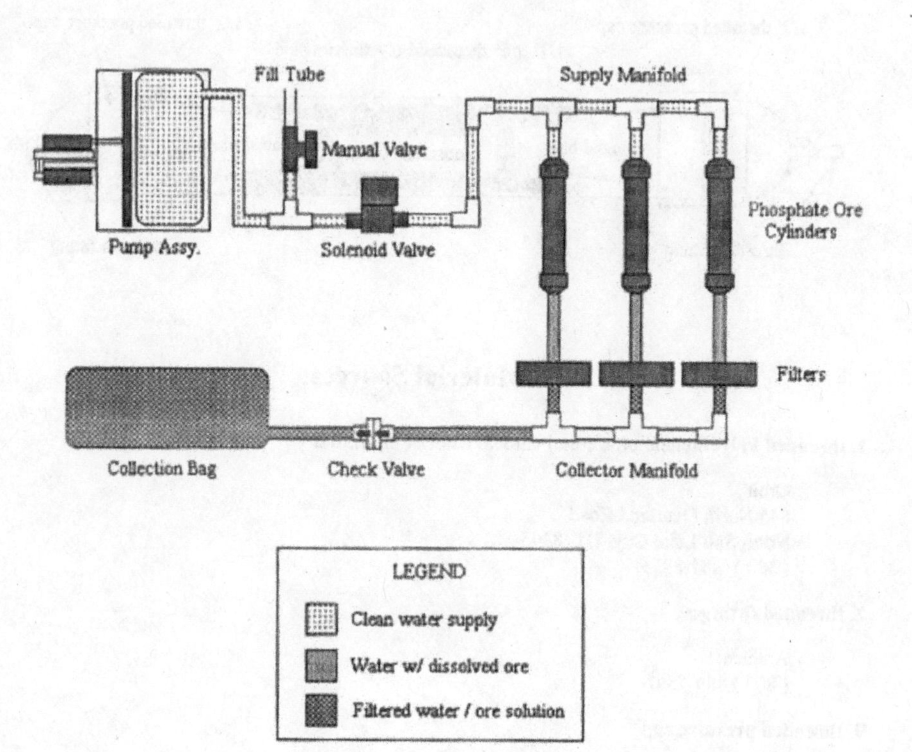

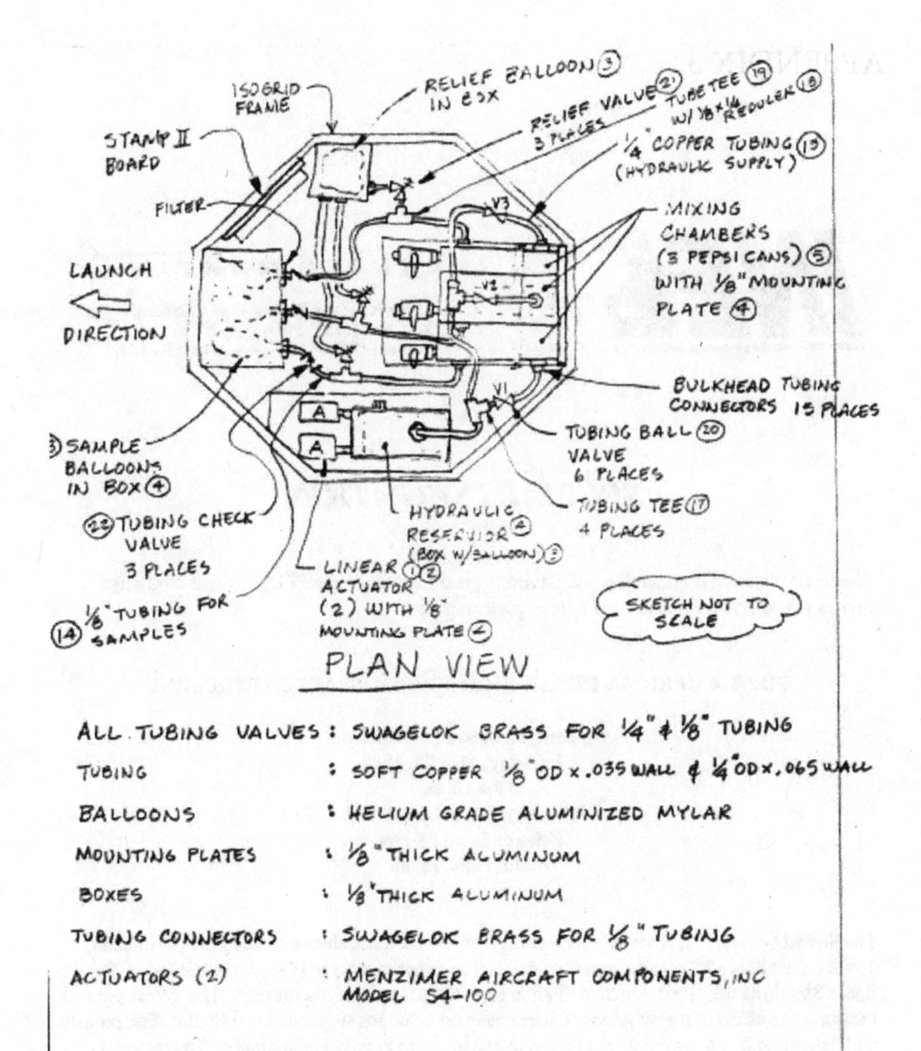
ISOGRID FRAME
RELIEF BALLOON (3)
IN BOX
RELIEF VALVE (21)
3 PLACES
TUBE TEE (19)
W/ 1/8"x1/4" REDUCER (18)
STAMP II BOARD
1/4" COPPER TUBING (15)
(HYDRAULIC SUPPLY)
FILTER
V3
MIXING CHAMBERS
(3 PEPSI CANS) (5)
WITH 1/8" MOUNTING PLATE (4)
LAUNCH
DIRECTION
V2
BULKHEAD TUBING CONNECTORS 15 PLACES
V1
A
A
TUBING BALL (20) VALVE 6 PLACES
(3) SAMPLE BALLOONS IN BOX (4)
TUBING TEE (17) 4 PLACES
(22) TUBING CHECK VALVE 3 PLACES
HYDRAULIC RESERVOIR (4)
(BOX W/BALLOON) (3)
LINEAR (1)(2) ACTUATOR
(2) WITH 1/8" MOUNTING PLATE (2)
1/8" TUBING FOR (14) SAMPLES
SKETCH NOT TO SCALE
PLAN VIEW
ALL TUBING VALVES : SWAGELOK BRASS FOR 1/4" & 1/8" TUBING
TUBING : SOFT COPPER 1/8" OD x .035 WALL & 1/4" OD x .065 WALL
BALLOONS : HELIUM GRADE ALUMINIZED MYLAR
MOUNTING PLATES : 1/8" THICK ALUMINUM
BOXES : 1/8" THICK ALUMINUM
TUBING CONNECTORS : SWAGELOK BRASS FOR 1/8" TUBING
ACTUATORS (2) : MENZIMER AIRCRAFT COMPONENTS, INC
MODEL S4-100

Original drawing of plan for Baby. Sketch by Ed Galindo

APPENDIX 3

LOCKHEED MARTIN

Lockheed Idaho Technologies Company
P. O. Box 1625 Idaho Falls, ID 83415
Telephone: (208) 526-0318 Facsimile (208) 526-1880

SPECIAL INVITATION

The Idaho National Engineering and Environmental Laboratory (INEEL) Institute cordially invites you and your students to a special presentation.

FIRST AMERICAN INDIAN HIGH SCHOOL SPACE EXPERIMENT

Student Technical Presentation
Thursday, May 22, 1997
10:00 a.m.
University Place Auditorium
Science Center Drive
Idaho Falls, Idaho

The Shoshone Bannock students successfully tested their space shuttle experiment in Houston, Texas on the KC135 "Vomit Comet" in April. The experiment is now scheduled to fly on the Space Shuttle as the "First American Indian High School Space Experiment". The experiment measures the effect of a zero gravity environment on solutions suspended in a liquid. The project will determine if space mining can be conducted in a zero gravity environment. The material used is Phosphate Ore which is mined at the Fort Hall Indian Reservation. The experiment will attempt to mix the ore with three different liquids, H2O, de-ionized H2O, and citric buffer in zero gravity and then attempt to filter out the liquids from the ore. The students have planned, designed and constructed the project as an after school project - INEEL Science Action Team.

The University of Idaho granted two undergraduate credits: ITED 410, Section 70, NASA Science and Technology.

We look forward to seeing you on the 22nd and supporting our neighbors at the Shoshone Bannock High School.

References

Alexander, Sonja and Anne Hutchison. 2000. "Indian Students to Use Mars 'Soil' to Grow Spuds in Space." *NASA News*. Houston, TX: Johnson Space Center. https://www.nasa.gov/centers/johnson/news/releases/1999_2001/h00-79.html. Accessed 2/7/2021.

Arnold, Mikayla, Sean N. Bennett, Melissa Christensen, Zachary Harmon, Camille Hatch, and Morghan Reid. 2010. *Transcultural Nursing*. http://freebooks.uvu.edu/NURS3400/index.php/ch09-shoshone-culture.html. Accessed 2/4/2021.

American Indian Report. 2003. *Indian Country's News Magazine,* 12-15.

"Astronauts Find No. 1 Reason to Become a Whiz at Art in Space." *Orlando Sentinel*, January 9, 2003. https://www.orlandosentinel.com/news/os-xpm-2003-01-29-0301290360-story.html. Accessed November 15, 2019.

Berkes, Fikret. 2012. *Sacred Ecology, 3rd Edition*. New York: Routledge.

Broncho, Lavern. 1995. Shoshone-Bannock tribal member interview.

Bureau of Indian Education. "Annual Performance Report for 2006–2007." https://www.bie.edu/cs/groups/xbie/documents/text/idc-008193.pdf. Accessed November 15, 2019.

Cajete, Gregory. 1994. *Look Toward the Mountain: An Ecology of Indigenous Education*. Skyland, N.C.: Kivaki Press.

Cajete, Gregory. 1999. *Igniting the Sparkle. An Indigenous Science Education Model.* Santa Fe, NM: Kivaki Press.

Cajete, Gregory. 2000. *Native Science: Natural Laws of Interdependence.* Santa Fe, NM: Clear Light Publishers.

Camera, Lauren. 2015. "Native American Students Left Behind," *US News and World Report*, November 16, 2015. https://www.usnews.com/news/articles/2015/11/06/Native-american-students-left-behind.

Chien, Philip. 2006. *Columbia: Final Voyage.* Göttingen, Germany: Copernicus Publishing.

Chung, Simmee. 2018. "Education Is Ceremony: Thinking *With* Stories of Indigenous Youth and Families." *LEARNing Landscapes* 11 (2): 93.

Cuch, Forrest. 2003. *A History of Utah's American Indians.* Logan, UT: Utah State University Press.

Deyhle, Donna and Karen Swisher. 1997. "Research in American Indian and Alaska Native Education: From Assimilation to Self-Determination." In *Review of Research in Education* 22 (1): 113–194. Accessed December 14, 2020, from http://www.jstor.org/stable/1167375.

Deloria, Vine Jr. 1978. "The Indian Student Amid American Inconsistencies." In *The Schooling of Native America*, edited by Thomas Thompson. (Washington DC: American Association of Colleges for Teacher Education).

Dewey, John. 1899. *The School and Society*. Chicago: University of Chicago Press.

Dunbar-Ortiz, Roxanne. 2014. *An Indigenous Peoples' History of the United States*. Boston: Beacon Press.

Fox, Sandra J. and Viola Lafontaine. 1999. "A Whole Language Approach to the Communication Skills: Nine Strategies for Teaching Language as an Integrated Whole to Indian Students." In *Teaching the Native American*, Fourth Edition, Hap Gilliland, 193–201. Dubuque, IA: Kendall/Hunt Publishing Company.

Frobisher, Martin, Ronald D. Hinsdill, Koby T. Crabtree, Clyde R. Goodheart. 1974. *Fundamentals of Microbiology*, 9th Edition. St. Louis, MO: W.B. Saunders Company.

Frost, Robert. 2002. *Robert Frost's Poems*. London: St. Martin's Paperbacks.

Galindo, Ed. 2017. "Mentoring American Indian Students with NASA." Public Lecture Series. https://digitalcommons.mtech.edu/public_lectures_mtech/94/.

Gilliland, Hap. 1999. *Teaching the Native American*. Dubuque, IA: Kendall/Hunt Publishers.

Guthridge, George. 1986. "Eskimos Solve the Future." *Analog: Science Fiction Science Fact* 106 (4): 67–172.

Harris, Bernard. 2019. "An Astronaut's Guide to Culturally Responsive Teaching." An EdSurge Podcast interview with Stephen Noonoo. https://soundcloud.com/edsurge/an-astronauts-guide-to-culturally-responsive-teaching.

Hawking, Stephen. (1988) 2017. *A Brief History of Time : From the Big Bang to Black Holes*. London: Bantam Books. Citation refers to 2017 edition.

Hawking, Stephen, and Roger Penrose. 2000. *The Nature of Space and Time*. Princeton (New Jersey): Princeton University Press.

Heady, Eleanor B. 1973. *Shoshone-Bannock Tribal Stories*. Fort Hall, ID: Shoshone-Bannock Tribal Library.

Howell, Elizabeth. 2018 (updated Oct. 12, 2021.) "International Space Station: Facts, History & Tracking." https://www.space.com/16748-international-space-station.html. Accessed 12/17/2020.

Idaho Centennial Commission, Native American Committee. 1990. *Idaho Indians: Tribal Histories*. Boise, Idaho: Idaho Centennial Commission.

Jenkins, Sally. 2012. "Why Are Jim Thorpe's Olympic Records Still Not Recognized?" *Smithsonian Magazine*. https://www.smithsonianmag.com/history/why-are-jim-thorpes-olympic-records-still-not-recognized-130986336/.

Juneau, Stan. 2013. *History and Foundation of American Indian Education*. https://jan.ucc.nau.edu/~jar/AIE/Juneau%20History.pdf.

Just, Rick. 2020. "The Lincoln Creek Day School." https://www.rickjust.com/blog/the-lincoln-creek-day-school.

Kirmayer, Laurence J. and Gail Guthrie Valaskakis, editors. 2009. *Healing Traditions: The Mental Health of Aboriginal Peoples in Canada.* Vancouver: UBC Press.

Kirkness, Verna J. and Ray Barnhardt. 2001. "First Nations and Higher Education: The Four R's—Respect, Relevance, Reciprocity, Responsibility." In *Knowledge Across Cultures: A Contribution to Dialogue Among Civilizations,* edited by Ruth Hayoe and Julia Pan. Comparative Education Research Centre, The University of Hong Kong. http://www.ankn.uaf.edu/IEW/winhec/FourRs2ndEd.html.

Kovach, M. Margaret. 2009. *Indigenous Methodologies: Characteristics, Conversations, and Contexts.* Ontario, Canada: University of Toronto Press.

Larsen, Natalie. 2019. "The Division of Fort Hall: Allotment and Land Loss." Intermountain Histories. Provost, Utah: Brigham Young University. https://www.intermountainhistories.org/items/show/203.

Larsen, Natalie. 2020. "Ranching and Success on Fort Hall: The Indian Stockman's Association." Intermountain Histories. Provost, Utah: Brigham Young University. https://www.intermountainhistories.org/items/show/205.

Marshall, Albert. Two Eyed Seeing. http://www.integrativescience.ca/Principles/TwoEyedSeeing. Accessed 12/13/2020.

Miller, Dorcas S. 1997. *Stars of the First People: Native American Star Myths and Constellations.* Portland, OR: Westwinds Press.

Monroe, Jean Guard and Ray Williamson. 1987. *They Dance in the Sky: Native American Star Myths.* Boston: Houghton Mifflin Company.

National Center for Educational Statistics. Fast Facts: Dropout Rates. https://nces.ed.gov/fastfacts/display.asp?id=16. Accessed November 15, 2019.

National Indian Council on Aging, Inc. (NICOA). 2019. https://www.nicoa.org/american-indian-veterans-have-highest-record-of-military-service/.

National Park Service. https://www.nps.gov/articles/000/dawes-act.htm. Accessed 1/12/2021.

Nebergall, William H., Frederic C. Schmidt, and Henry F. Holtzclaw, Jr. 1973. *College Chemistry with Qualitative Analysis*, 4th Edition. Urbana, IL: University of Illinois.

One Spot, Mary. 1998. In *The Ways of my Grandmothers*, Beverly Hungry Wolf. New York City: William Morrow Publishing Paperbacks.

Prucha, Francis Paul. 1984. *The Great Father: The United States Government and the American Indians*, Volume I. Lincoln: University of Nebraska Press.

Ramesh, Kulasekaran and Dendi Damodar Reddy. 2011. "Zeolites and Their Potential Uses in Agriculture." *Advances in Agronomy* 113. New York: Elsevier Inc.

Reyhner, Jon and Jeanne Eder. 2017. *American Indian Education: A History*, 2nd Edition. Norman, OK: University of Oklahoma Press.

Scott, Ridley. 2015. *The Martian*. United States: Twentieth Century Fox.

Shoshone-Bannock Tribes. www.sbtribes.com.

Sister Sky. 2019. https://sistersky.com/blogs/sister-sky/the-significance-of-hair-in-Native-american-culture.

Slapin, Beverly and Dorothy Seale, editors. 1998. *Through Indian Eyes: The Native Experience in Books for Children*. 4th Edition (Contemporary American Indian Issues No. 7). Los Angeles: University of Southern California American Indian Studies Center.

Smith, Joseph. 1830. *Book of Mormon*. https://www.josephsmithpapers.org/paper-summary/book-of-mormon-1830/1.

Space.com. 2020. "The Mercury 7 Astronauts: NASA's First Space Travelers." https://www.space.com/25398-nasa-mercury-7-astronauts-first-americans-in-space.html. Accessed 12/17/2020.

True West. 2021. *True West Ultimate Historical Guide*, 4th edition. Cave Creek, AZ: Two Roads West.

USGS.edu. 2018. The Water Cycle for Schools. https://www.usgs.gov/special-topics/water-science-school/science/water-cycle-schools.

Waldman, Carl. 2006. *Encyclopedia of Native American Tribes*. London, UK: Checkmark Books. (Found on https://www.newworldencyclopedia.org/entry/yaqui. Accessed 12/21/2020.)

Wasco legend (epigraph, chapter 3). https://www.firstpeople.us/FP-Html-Legends/Coyote-Places-The-Stars-Wasco.html. Accessed 2/1/2021.

Wilson, Shawn. 2008. *Research is Ceremony: Indigenous Research Methods*. Nova Scotia, Canada: Fernwood Publishing.

Worldometer. 2019. "China Population—Worldometers." https://www worldometers.info/world-population/china-population/.

Additional Reading

I am including a few friends from my own reading list that you may be interested in as well.

Boyd, Doug. 1974. *Rolling Thunder*. London: Delta Books.

Boyer, Paul, editor. 2010. *Ancient Wisdom, Modern Science: The Integration of Native Knowledge in Math and Science at Tribally Controlled Colleges and Universities.* Pablo, MT: Salish Kootenai College Press.

Cleary, Linda Miller and Thomas D. Peacock. 1998. *Collected Wisdom: American Indian Education.* Boston: Allyn and Bacon.

Eastman, Charles (Ohiyesa). 2010. *Living in Two Worlds the American Indian Experience.* Bloomington, IN: World Wisdom.

Gladwell, Malcolm. 2007. *Outliers. The Story of Success.* Boston: Little, Brown and Company.

Kimmerer, Robin Wall. 2003. *Braiding Sweetgrass: Indigenous Wisdom, Scientific Knowledge, and the Teaching of Plants*. Minneapolis: Milkweed Editions.

Lambert, Lori. 2014. *Research for Indigenous Survival: Indigenous Research Methodologies for the Behavioral Sciences*. Pablo, MT: Salish Kootenai College Press.

Maryboy, Nancy C. and David Begay. 2010. *Sharing the Skies: Navajo Astronomy, A Cross-Cultural View.* Tucson, AZ: Rio Nuevo Publishers.

Mihesuah, Devon A. 2000. *The Roads of my Relations*. Tucson: University of Arizona Press.

Mitchell, Sherri [Weh'na'Kwasset (She Who Brings the Light)]. 2018. *Sacred Instructions. Indigenous Wisdom for Living Spirit-Based Change*. Berkeley, California: North Atlantic Books.

National Museum of the American Indian. 2007. *Do All Indians Live in Tipis?* Questions and Answers from the National Museum of the American Indian. (Introduction by Wilma Mankiller. Foreword by Rick West). Smithsonian Books.

Palmer, William. 1978. *Why the North Stands Still and Other Indian Legends*. Springdale, UT: Zion Natural History Association.

Pavel, D. Michael (CHiXapkaid) and Ella Inglebret. 2007. *American Indian and Alaska Native Students Guide to College Success*. Westport, CT: Greenwood Press.

Peat, F. David. 2005. *Blackfoot Physics: A Journey into the Native American Universe.* Boston and York Beach, ME: Weiser Books.

Weir, Andy. 2013. *The Martian*. New York: Broadway Books.